# MODAL APPROXIMATIONS

*Theory and an Application to Reactor Physics*

# MODAL APPROXIMATIONS

## *Theory and an Application to Reactor Physics*

---

WESTON M. STACEY, JR.

RESEARCH MONOGRAPH NO. 41
THE M.I.T. PRESS, CAMBRIDGE, MASSACHUSETTS

*Set in Monotype Times New Roman and*
*printed and bound in the United States of America*
*by The Riverside Press, Cambridge, Massachusetts*



*Library of Congress catalog card number: 67-16498*

# Foreword

This the forty-first volume in the M.I.T. Research Monograph Series published by the M.I.T. Press. The objective of this series is to contribute to the professional literature a number of significant pieces of research, larger in scope than journal articles but normally less ambitious than finished books. We believe that such studies deserve a wider circulation than can be accomplished by informal channels, and we hope that this form of publication will make them readily accessible to research organizations, libraries, and independent workers.

HOWARD W. JOHNSON

# Acknowledgments

The author wishes to express appreciation for the encouragement and stimulation provided during the course of this work by his thesis supervisors, Dr. Irving Kaplan and Dr. Kent F. Hansen.

The author is grateful to the Atomic Energy Commission, the Fannie and John Hertz Foundation, and the Knolls Atomic Power Laboratory for financial support.

# Contents

# 1. Introduction

Practical solutions are obtained for many problems in engineering by replacing the rigorous equation that describes the problem by a more mathematically tractable set of approximate equations with a reduced number of independent variables. A variety of means have been employed to generate the approximate set of equations; moments approximations and polynomial expansions are familiar examples. In many cases, the nature of the approximations involved and the physical or mathematical significance of the resulting set of approximate equations is obscure. Moreover, the "best" choice among several alternate means of generating approximate equations is not apparent. The semidirect variational method has been suggested[33,58] as a systematic procedure for generating approximate equations that would resolve this latter quandary, although equations resulting from the variational procedure are "best" in a special sense.

One of the purposes of this monograph is to codify a flexible procedure for generating a set of approximate equations with a reduced number of independent variables. The physical or mathematical significance of the approximate equations is clear from the means whereby they are obtained, and the nature of the approximation involved is evident. This procedure can be systematically applied to obtain "best" sets of approximate equations in the sense that the solution of these approximate equations possesses a certain minimal error property. This general procedure for generating approximate equations is referred to as the *modal expansion method.*

The modal expansion method is well suited for developing approximate calculational models. The nature of the approximations involved and the implied significance of the results are apparent, and in addition, general information about the nature of the solution may be incorporated into the calculational model via the choice of expansion modes, thus enhancing the model.

Calculational models for the stationary neutron space-energy distribution are derived from the modal expansion method. In addition, it is found that many of the standard approximations of reactor physics may be obtained by a modal expansion. Thus, the modal expansion method provides a general framework within which alternative approximations may be compared. In particular, the conventional multigroup–finite difference treatment is shown to be a special case of a modal expansion, and is qualitatively compared with more general modal expansions.

Approximate calculational models are developed for the stationary neutron distribution in fast-reactor assemblies. This application was motivated by the importance of fast reactors in the United States nuclear power program[31] and by the paucity of simple models for fast-reactor calculations.

The conventional multigroup–finite difference calculation of the neutron distribution is limited by the storage capacity of modern digital computers. Because of elastic scattering resonances in intermediate atomic mass nuclei, a large number of groups are required to represent the fine structure in the neutron spectrum (one code employs almost 700 groups). This limits the representation of the spatial and angular dependence, and results in a complex calculation that tends to obscure the physical aspects of the problem. Several calculations of representative fast-reactor assemblies indicate that modal approximations may be a feasible means of alleviating certain of these disadvantages associated with the standard multigroup–finite difference calculation.

In order to obtain general information that may be advantageously incorporated into the modal calculational model, a simple means of generating infinite medium spectra and importance functions is desirable. Analytical expressions are developed for these quantities, both by solving the appropriate equations by the method of successive approximations and by constructing a solution directly from physical arguments.

Thus, the main purposes of this monograph are to codify the modal expansion method, to apply this method to obtain simple calculational models for fast-reactor assemblies, and to develop analytical expressions for the spatially independent neutron spectrum and importance function.

# 2. Modal Expansion Method

Modal expansion techniques may be characterized as those procedures by which the number of independent variables in a problem is reduced by means of an expansion in known functions of one or more of the independent variables. The choice of the expansion functions, or *modes*, and the determination of the coefficients with which the modes are combined to obtain a solution are central issues of the modal expansion method.

In this chapter the general formalism is codified, criteria are developed for selecting modes and determining the combining coefficients, and the inclusive character of the modal expansion method is illustrated by citing numerous approximate techniques which are special cases.

## 2.1 General Formalism

Consider a linear inhomogeneous equation in $n$ independent variables $(x_1, \cdots, x_n)$,

$$H(x_1, \cdots, x_n)\phi(x_1, \cdots, x_n) - f(x_1, \cdots, x_n) = 0, \tag{2.1}$$

where $H$ is a linear operator, $\phi$ is the solution, and $f$ is an inhomogeneous "source." A modal expansion of the solution, relative to the independent variable $x_1$, is defined by

$$\phi(x_1, \cdots, x_n) = \sum_{i=1}^{N} \psi_i(x_1)\rho_i(x_2, \cdots, x_n), \tag{2.2}$$

where the $\psi_i$ are specified functions of $x_1$.

The combining coefficients $\rho_i$ are determined by imposing the requirement that the modal expansion satisfies the governing equation in a weighted integral sense, for $N$ different weighting operations. This is effected by substituting Equation 2.2 into Equation 2.1, operating with an arbitrary weighting operator $\alpha_j(x_1,\cdots,x_n)$, and integrating over $x_1$. This procedure is repeated to obtain $N$ coupled equations,

$$\int dx_1\alpha_j(x_1,\cdots,x_n)\left[H(x_1,\cdots,x_n)\sum_{i=1}^{N}\psi_i(x_1)\rho_i(x_2,\cdots,x_n) - f(x_1,\cdots,x_n)\right] = 0, \qquad j = 1,\cdots,N, \tag{2.3}$$

in $n-1$ independent variables $(x_2,\cdots,x_n)$, which determine the combining coefficients $\rho_i$. These equations must be linearly independent. Equations 2.3 may be written as a matrix equation,

$$\mathsf{H}(x_2,\cdots,x_n)\cdot\mathsf{\rho}(x_2,\cdots,x_n) - \mathsf{f}(x_2,\cdots,x_n) = 0, \tag{2.4}$$

where the matrix elements are given by

$$H^{ji}(x_2,\cdots,x_n) \equiv \int dx_1\alpha_j(x_1,\cdots,x_n)H(x_1,\cdots,x_n)\psi_i(x_1),$$

$$f^j(x_2,\cdots,x_n) \equiv \int dx_1\alpha_j(x_1,\cdots,x_n)f(x_1,\cdots,x_n),$$

and $\mathsf{\rho}$ is the column matrix formed by the $\rho_i$. In general, Equation 2.4 is a matrix operator equation; that is, the matrix elements $H^{ji}$ are operators.

A modal expansion may be applied to one, several, or all of the independent variables; the results are formally identical to those obtained above if $x_1$ is generalized to represent all such independent variables. When a modal expansion is made for all of the independent variables, Equation 2.4 represents a system of algebraic equations. The terminology "direct" will be employed to distinguish those techniques for which Equation 2.4 is an algebraic matrix equation from those for which it is a nonalgebraic operator equation. The latter will be referred to as "semidirect" techniques.

When the weighting operators are functions of $x_1$ only, the matrix elements $H^{ji}(x_2,\cdots,x_n)$ have the same functional dependence upon $x_2,\cdots,x_n$ as does the original operator, $H(x_1,\cdots,x_n)$.† Consequently, Equation 2.4 has the same functional form that Equation 2.1 would have if the $x_1$ dependence was suppressed. This formal analogy suggests that solution techniques appropriate for the latter, which is a scalar equation, can be adapted to solve the former, which is a matrix equation.

† This property also obtains for certain other weighting operators.

## 2.2 Modes and Weighting Operators

The modes $\psi_i(x_1)$ constitute the basis of a linear vector space. The procedure of substituting Equation 2.2 into Equation 2.1, operating thereupon with $\alpha_j$, and integrating over $x_1$ is equivalent to projecting Equation 2.1 onto the vector space defined by the modes $\psi_i(x_1)$. The nature of the projection depends upon the weighting operator $\alpha_j$. Thus, Equations 2.3 are $N$ different projections of Equation 2.1 onto the vector space defined by the $\psi_i(x_1)$, and constitute a representation of Equation 2.1 upon that space. There are three cases (at least) for which this representation is exact; that is, the solutions of Equations 2.3 may be used to construct an exact solution of Equation 2.1 of the form of Equation 2.2.

Equations 2.3 are exact when the modes $\psi_i$ constitute a complete set of linearly independent functions on some interval of $x_1$, and the weighting operators are chosen as the functions which are biorthonormal to the modes. Using the notation of Dirac,[16] these two conditions may be written (see Appendix A for a review of linear vector space notation)

$$1 \equiv \sum_{i=1}^{N} |\psi_i(x_1)\rangle\langle\alpha_i(x_1')|, \tag{2.5a}$$

and

$$\delta_{ij} = \langle\alpha_j(x_1) \mid \psi_i(x_1)\rangle. \tag{2.5b}$$

When Identity 2.5$a$ is substituted into Equation 2.1, the following exact equation obtains,

$$\sum_{ij=1}^{N} |\psi_i(x_1)\rangle\langle\alpha_i(x_1')|H(x_1',\cdots,x_n)|\psi_j(x_1')\rangle\langle\alpha_j(x_1'')|\phi(x_1'',\cdots,x_n)\rangle - \sum_{i=1}^{N} |\psi_i(x_1)\rangle\langle\alpha_i(x_1')|f(x_1',\cdots,x_n)\rangle = 0. \tag{2.6a}$$

The inner product is with respect to the independent variable $x_1$, only.

Because the $|\psi_i\rangle$ are linearly independent, Equation 2.6$a$ is equivalent to the $N$ exact relations,

$$\sum_{j=1}^{N} \langle\alpha_i(x_1')|H(x_1',\cdots,x_n)|\psi_j(x_1')\rangle\langle\alpha_j(x_1'')|\phi(x_1'',\cdots,x_n)\rangle - \langle\alpha_i(x_1')|f(x_1',\cdots,x_n)\rangle = 0, \qquad i = 1,\cdots,N,$$

which are formally identical to Equations 2.3, when it is noted that $\langle\alpha_j|\phi\rangle$ is equivalent to $\rho_j$.

When the modes do not constitute a complete set, Equations 2.3 may still be exact under certain conditions. Consider a complete set of functions $\psi_i$, $i = 1,\cdots, N$ and an incomplete subset $\psi_i$, $i = 1,\cdots, M < N$. The exact representation, Equation 2.6*a*, may be rearranged,

$$\begin{aligned}\sum_{ij=1}^{M} |\psi_i\rangle\langle\alpha_i|H|\psi_j\rangle\langle\alpha_j|\phi\rangle - \sum_{i=1}^{M} |\psi_i\rangle\langle\alpha_i|f\rangle \\ = \sum_{i=M+1}^{N} |\psi_i\rangle\langle\alpha_i|f\rangle - \sum_{i=1}^{N}\sum_{j=M+1}^{N} |\psi_i\rangle\langle\alpha_i|H|\psi_j\rangle\langle\alpha_j|\phi\rangle \\ - \sum_{i=M+1}^{N}\sum_{j=1}^{M} |\psi_i\rangle\langle\alpha_i|H|\psi_j\rangle\langle\alpha_j|\phi\rangle.\end{aligned} \tag{2.6b}$$

This reduces to Equations 2.3 if the right side vanishes. Sufficient conditions for the vanishing of the first two terms on the right side of Equation 2.6*b* are

$$\langle\alpha_j(x_1)|f(x_1,\cdots, x_n)\rangle = \langle\alpha_j(x_1)|\phi(x_1,\cdots, x_n)\rangle = 0, \qquad j = M + 1,\cdots, N. \tag{2.7a}$$

The third term on the right side of Equation 2.6*b* vanishes if

$$\langle\alpha_j(x_1)|H(x_1,\cdots, x_n)|\psi_i(x_1)\rangle = 0, \qquad j = M + 1,\cdots, N, \tag{2.7b}$$

or if the $|\psi_i\rangle$ are eigenfunctions of $H$ in the sense

$$H(x_1,\cdots,x_n)\psi_i(x_1)\rho(x_2,\cdots, x_n) = \lambda_i\psi_i(x_1)K(x_2,\cdots, x_n)\rho(x_2,\cdots, x_n), \tag{2.7c}$$

where $\rho$ is a function and $K$ is a linear operator.

When Equations 2.7*a* and 2.7*b* obtain, Equation 2.6*b* becomes, because of the linear independence of the $\psi_i$, an exact set of $M$ equations,

$$\sum_{j=1}^{M} \langle\alpha_i(x_1)|H(x_1,\cdots, x_n)|\psi_j(x_1)\rangle\langle\alpha_j(x_1')|\phi(x_1',\cdots, x_n)\rangle - \langle\alpha_i(x_1)|f(x_1,\cdots, x_n)\rangle = 0, \qquad i = 1,\cdots, M,$$

which are formally identical with Equations 2.3. (The dummy number $N$ of Equation 2.3 is equivalent to $M$ of this equation.)

On the other hand, when Equations 2.7*a* and 2.7*c* are satisfied, Equation 2.6*b* becomes

$$\lambda_j K(x_2,\cdots, x_n)\langle\alpha_j(x_1)|\phi(x_1,\cdots, x_n)\rangle - \langle\alpha_j(x_1)|f(x_1,\cdots, x_n)\rangle = 0, \qquad j = 1,\cdots, M,$$

which also are identical with Equations 2.3, subject to Equation 2.7*c*.

Weighting functions that satisfy Equation 2.5*b* project out the component of a function along a single $\psi_i$. Thus, Equations 2.7*a* are requirements that $f$ and $\phi$ have no component along $\psi_i$, $i = M + 1, \cdots, N$.

The set of Equations 2.7*b* is a requirement that $H\psi_j$, $j = 1, \cdots, M$, has no component along $\psi_i$, $i = M + 1, \cdots, N$.

Consequently, Equations 2.3 are exact when the modes form a complete set and the weighting operators are the complete set of functions which are biorthogonal to the modes; that is, the modes and weighting functions satisfy Equations 2.5. In addition, Equations 2.3 may be exact when the modes do not constitute a complete set if the following conditions are fulfilled:

1. The $\psi_i(x_1)$ satisfy either Equations 2.7*b* or 2.7*c*.
2. The $\alpha_j(x_1)$ are functions biorthogonal to $\psi_i(x_1)$; that is, satisfy Equation 2.5*b* on an appropriate interval of $x_1$.
3. The $\psi_i(x_1)$, $i = 1, \cdots, M$, form a complete subset with respect to $\phi$ and $f$; that is, Equations 2.7*a* are satisfied.

In general, it is impractical, if not impossible, to work with a complete set of modes. However, it is possible to use physical insight as to the nature of the solution, and knowledge of the source, to select modes such that Conditions 3 and 1 are approximated. When the $\alpha_j$ are chosen to satisfy Condition 3, Equation 2.4 approaches an exact equation in the limit that the modes and weighting functions satisfy the three conditions exactly.

Many practical applications of modal expansion techniques may have as their objective an approximate solution of Equation 2.1. In these instances, the above considerations regarding the choice of the weighting operators $\alpha_j$ may be replaced by considerations which are more appropriate for approximate solutions. Condition 3 would remain a valid criterion for selection of modes, however. The exact solution to Equation 2.1 may be written

$$\phi_0(x_1, \cdots, x_n) = \sum_{i=1}^{N} \psi_i(x_1)\rho_i(x_2, \cdots, x_n) + \delta\phi(x_1, \cdots, x_n), \tag{2.8}$$

where the $\rho_i$ are determined by solving Equations 2.3. For a given set of modes $\psi_i(x_1)$ the choice of weighting operators $\alpha_j(x_1, \cdots, x_n)$ affects the $\rho_i$ that are determined from Equations 2.3 and, consequently, affects the error, $\delta\phi$. Thus, the weighting operators may be chosen to minimize the error, *in some sense*. The resulting Equations 2.3 are then the "best" approximate equations in that their solution ($\rho_i$) leads to an approximate solution of Equation 2.1 of the form of Equation 2.2, which has the property that $\delta\phi$ is minimal, in that sense. Several possibilities are examined.

(*a*) *Minimize the error in the system equation in a least-squares sense.* The error $\delta\phi$ in the solution results in an error $H\delta\phi$ in the system equation. The requirement that this latter be a minimum in a least-squares sense is written

$$\int dx_n \cdots \int dx_1 |H\delta\phi|^2 = \text{minimum}. \tag{2.9}$$

When Equation 2.8 is used for $\delta\phi$, and the fact that $\phi_0$ satisfies Equation 2.1 identically is accounted for, Equation 2.9 becomes

$$\int dx_n \cdots \int dx_1 \left| H(x_1, \cdots, x_n) \sum_{i=1}^{N} \psi_i(x_1)\rho_i(x_2, \cdots, x_n) - f(x_1, \cdots x_n) \right|^2 = \text{minimum}.$$

For this integral to be a minimum, its variation must vanish through first order for arbitrary variations of each of the $\rho_i$ about the values that satisfy Equation 2.9. From this requirement result the $N$ equations

$$\int dx_1 \psi_j(x_1) H^*(x_1, \cdots, x_n) \left[ H(x_1, \cdots, x_n) \sum_{i=1}^{N} \psi_i(x_1)\rho_i(x_2, \cdots, x_n) - f(x_1, \cdots, x_n) \right] = 0, \qquad j = 1, \cdots, N. \tag{2.10}$$

Comparison of these equations with Equations 2.3 indicates that

$$\alpha_j(x_1, \cdots, x_n) = \psi_j(x_1) H^*(x_1, \cdots, x_n), \tag{2.11}$$

is the weighting operator which minimizes the error in the system equation in a least-squares sense. Here $H^*$ is the operator adjoint to $H$ in the sense,[47]

$$\int dx_n \cdots \int dx_1 u(x_1, \cdots, x_n) H(x_1, \cdots, x_n) \phi(x_1, \cdots, x_n)$$
$$\equiv \int dx_n \cdots \int dx_1 \phi(x_1, \cdots, x_n) H^*(x_1, \cdots, x_n) u(x_1, \cdots, x_n),$$

where $u$ and $\phi$ are arbitrary functions and the bilinear concomitant is assumed to vanish.

Equations 2.10 may be obtained by an extension of the method of least squares[42] to a multivariate problem, or, equivalently, by a semidirect application of the least-squares variational principle.[4] In this respect, these methods may be considered as special cases of the modal expansion method.

(*b*) *Eliminate the net effect of the error on integral physical parameters.*

The term $H\delta\phi$ may be considered as a pseudo-source. If $G_a(x_1,\cdots, x_n)$ is the kernel which specifies the effect of a unit source at $(x_1,\cdots, x_n)$ on some physical parameter, the requirement

$$\int dx_1 G_a(x_1,\cdots, x_n)H(x_1,\cdots, x_n)\delta\phi(x_1,\cdots, x_n) = 0, \qquad (2.12)$$

ensures that the net effect of the error $\delta\phi$ on the physical parameter associated with $G_a$ vanishes; that is, the physical parameter is calculated correctly.

When the physical parameter $A$ is given by an integral of the form

$$A = \int dx_n \cdots \int dx_1 \Sigma_a(x_1,\cdots, x_n)\phi(x_1,\cdots, x_n), \qquad (2.13)$$

the appropriate kernel $G_a$ is obtained by solving the equation adjoint to Equation 2.1 with $\Sigma_a$ as the inhomogeneous term,

$$H^*(x_1,\cdots, x_n)G_a(x_1,\cdots, x_n) - \Sigma_a(x_1,\cdots, x_n) = 0. \qquad (2.14)$$

The proof of this statement follows from multiplying Equation 2.1 by $G_a$ and Equation 2.14 by $\phi$, subtracting, and integrating.

$$\int dx_n \cdots \int dx_1 [G_a H\phi - G_a f - \phi H^* G_a + \phi\Sigma_a] = 0.$$

The first and third terms cancel by definition of $H^*$, assuming the bilinear concomitant of the operator $H$ vanishes. Thus,

$$\int dx_n \cdots \int dx_1 G_a(x_1,\cdots, x_n) f(x_1,\cdots, x_n)$$
$$= \int dx_n \cdots \int dx_1 \Sigma_a(x_1,\cdots, x_n)\phi(x_1,\cdots, x_n). \qquad (2.15)$$

From Equation 2.15 it is apparent that $G_a$ is the kernel which describes the effect of a unit source at $(x_1,\cdots, x_n)$ on the integral physical parameter $A$.

When $\delta\phi$ is replaced by its equivalent from Equation 2.8, Equation 2.12 becomes

$$\int dx_1 G_a(x_1,\cdots, x_n)\Big[H(x_1,\cdots, x_n)\sum_{i=1}^{N} \psi_i(x_1)\rho_i(x_2,\cdots, x_n) - f(x_1,\cdots, x_n)\Big] = 0.$$

Consequently,

$$\alpha_j(x_1,\cdots, x_n) = G_a(x_1,\cdots, x_n), \qquad (2.16)$$

is the weighting function that eliminates the net effect of the error $\delta\phi$ on the physical parameter $A$. The kernel $G_a$ is generally as difficult to obtain as the solution to Equation 2.1. However, the foregoing discussion illustrates the type of functions that should be sought in constructing approximate weighting functions.

(*c*) *Eliminate the error in the system equation at selected values of* $x_1$. The condition that the system equation be satisfied at $N$ discrete points,

$$H(x_{1,i}, x_2, \cdots, x_n)\delta\phi(x_{1,i}, x_2, \cdots, x_n) = 0, \qquad i = 1, \cdots, N,$$

may be fulfilled by choosing

$$\alpha_i(x_1, \cdots, x_n) = \delta(x_1 - x_{1,i}) \tag{2.17}$$

in Equations 2.3.

(*d*) *Eliminate the net error in the system equation over selected intervals of* $x_1$. The requirement,

$$\int_{x_{1,i-1}}^{x_{1,i}} dx_1 H(x_1, \cdots, x_n)\delta\phi(x_1, \cdots, x_n) = 0, \qquad i = 1, \cdots, N, \tag{2.18}$$

can be satisfied by selecting

$$\alpha_i(x_1) = -U(x_{1,i}) + U(x_{1,i-1}), \qquad U(x_{1,i}) = \begin{cases} 1, & x_1 \geqslant x_{1,i}, \\ 0, & x_1 < x_{1,i}. \end{cases} \tag{2.19}$$

When the system equation, Equation 2.1, is a balance equation, this choice of weighting function has the physical interpretation of requiring that the balance condition is satisfied in an integral sense over the interval $x_{1,i-1} \leqslant x_1 \leqslant x_{1,i}$.

(*e*) *Minimize the error in the solution in a least-squares sense.* The "best" approximation to the solution of Equation 2.1 of the form of Equation 2.2, in a mean sense, is found by requiring

$$\int dx_n \cdots \int dx_1 \, |\delta\phi(x_1, \cdots, x_n)|^2 = \text{minimum}, \tag{2.20}$$

or,

$$\int dx_n \cdots \int dx_1 \left| \phi_0(x_1, \cdots, x_n) - \sum_{i=1}^{N} \psi_i(x_1)\rho_i(x_2, \cdots, x_n) \right|^2 = \text{minimum}$$

For a given choice of $\psi_i$, this integral must be stationary with respect to arbitrary variations of the $\rho_i$ about the values which satisfy the minimum condition. This condition is equivalent to

$$\int dx_1 \bar{\psi}_i(x_1)\left[\phi_0(x_1,\cdots,x_n) - \sum_{j=1}^{N} \psi_j(x_1)\rho_j(x_2,\cdots,x_n)\right] = 0, \qquad i = 1,\cdots,N.$$

$\bar{\psi}_i$ is the function which is conjugate to $\psi_i$. The exact solution of Equation 2.1 may be written

$$\phi_0(x_1,\cdots,x_n) = H^{-1}(x_1,\cdots,x_n)f(x_1,\cdots,x_n).$$

Thus, the preceding equations become

$$\int dx_1 \bar{\psi}_i(x_1)H^{-1}(x_1,\cdots,x_n)\left[H(x_1,\cdots,x_n)\sum_{j=1}^{N}\psi_j(x_1)\rho_j(x_2,\cdots,x_n) - f(x_1,\cdots,x_n)\right] = 0, \qquad i = 1,\cdots,N. \tag{2.21}$$

Comparison with Equations 2.3 indicates that the choice

$$\alpha_i(x_1,\cdots,x_n) = \bar{\psi}_i(x_1)H^{-1}(x_1,\cdots,x_n) \tag{2.22}$$

minimizes the error in the solution in a least-squares sense.

When $\bar{\psi}_i$ and $\psi_i$ are biorthonormal, Equation 2.21 reduces to the familiar definition of the Fourier coefficient

$$\rho_i(x_2,\cdots,x_n) = \int dx_1 \bar{\psi}_i(x_1)H^{-1}(x_1,\cdots,x_n)f(x_1,\cdots,x_n) = \int dx_1 \bar{\psi}_i(x_1)\phi_0(x_1,\cdots,x_n).$$

If the operation $H^{-1}f$ could be performed, it would not be necessary to resort to modal expansion, as the solution would be known. However, it may be possible to construct an approximation to $H^{-1}$ which can be used in Equations 2.21 and to develop the solution iteratively.

(*f*) *Eliminate the error in the solution at N values of* $x_1$. The requirement

$$\delta\phi(x_{1,j}, x_2,\cdots,x_n) = 0, \qquad j = 1,\cdots,N, \tag{2.23}$$

or equivalently,

$$H^{-1}(x_{1,j},\cdots,x_n)\left[f(x_{1,j}, x_2,\cdots,x_n) - H(x_{1,j},\cdots,x_n)\sum_{i=1}^{N}\psi_i(x_{1,j})\rho_i(x_2,\cdots,x_n)\right] = 0, \qquad j = 1,\cdots,N,$$

can be imposed on Equations 2.3 by selecting the weighting operators as

$$\alpha_j(x_1,\cdots,x_n) = \delta(x_1 - x_{1,j})H^{-1}(x_1,\cdots,x_n). \tag{2.24}$$

(*g*) *Conserve the first N moments with respect to $x_1$ of the system equation.* Because $H\delta\phi$ can be considered a pseudo-source, the requirements

$$\int dx_1 x_1{}^j H(x_1,\cdots,x_n)\delta\phi(x_1,\cdots,x_n) = 0, \qquad j = 0, 1,\cdots, N-1, \tag{2.25}$$

ensure that the first $N$ moments of this source vanish. Equations 2.25 can be satisfied by selecting

$$\alpha_j(x_1,\cdots,x_n) = x_1{}^j, \tag{2.26}$$

in Equations 2.3.

(*h*) *Conserve the first N moments with respect to $x_1$ of the solution.* Equations 2.3 will correctly predict the first $N$ moments of the solution; that is,

$$\int dx_1 x_1{}^j \delta\phi(x_1,\cdots,x_n) = 0, \qquad j = 0, 1,\cdots, N-1, \tag{2.27}$$

if the weighting operators are

$$\alpha_j(x_1,\cdots,x_n) = x_1{}^j H^{-1}(x_1,\cdots,x_n). \tag{2.28}$$

(*i*) *Render a variational functional stationary.* A variational functional $L(\phi^*, \phi)$ has the property of stationarity with respect to variations $\delta\phi$ and $\delta\phi^*$ about the exact solutions $\phi_0$ and $\phi_0{}^*$ of

$$H(x_1,\cdots,x_n)\phi_0(x_1,\cdots,x_n) - f(x_1,\cdots,x_n) = 0, \tag{2.29a}$$

and the adjoint equation

$$H^*(x_1,\cdots,x_n)\phi_0{}^*(x_1,\cdots,x_n) - f^*(x_1,\cdots,x_n) = 0. \tag{2.29b}$$

The Lagrangian functional, which includes as special cases many common variational functions, is[58]

$$L(\phi^*, \phi) = \int dx_n \cdots \int dx_1[f\phi^* + f^*\phi - \phi^* H\phi]. \tag{2.30}$$

The stationarity property follows from

$$\begin{aligned}\delta L &\equiv L(\phi_0{}^* + \delta\phi^*, \phi_0 + \delta\phi) - L(\phi_0{}^*, \phi_0) \\ &= \int dx_n \cdots \int dx_1[\delta\phi^*\{f - H\phi_0\} + \{f^* - \phi_0{}^* H\}\delta\phi - \delta\phi^* H\delta\phi].\end{aligned}$$

The first and second terms vanish by virtue of Equations 2.29 and the definition of the adjoint operator. This reduces to the stationarity property

$$\delta L = 0 + \text{order}\,(\delta\phi^* \cdot \delta\phi). \tag{2.31}$$

The Lagrangian functional may be applied in a semidirect fashion to obtain equations similar to Equations 2.3,[58] and thus, to specify the nature of the weighting functions for the modal expansion method. The solutions to Equations 2.29 are expanded in known functions $\psi_i(x_1)$ and $\psi_i^*(x_1)$,

$$\begin{aligned} \phi(x_1,\cdots,x_n) &= \sum_{i=1}^{N} \psi_i(x_1)\rho_i(x_2,\cdots,x_n), \\ \phi^*(x_1,\cdots,x_n) &= \sum_{i=1}^{N} \psi_i^*(x_1)\rho_i^*(x_2,\cdots,x_n). \end{aligned} \tag{2.32}$$

These expansions are substituted into Equation 2.30, and the functional is required to be stationary with respect to arbitrary variations of $\rho_i^*$ and $\rho_i$. This results in a set of $N$ equations for the $\rho_i$,

$$\int dx_1 \psi_j^*(x_1)\left[H(x_1,\cdots,x_n)\sum_{i=1}^{N}\psi_i(x_1)\rho_i(x_2,\cdots,x_n) - f(x_1,\cdots,x_n)\right] = 0, \qquad j = 1,\cdots,N, \tag{2.33}$$

and another set of $N$ equations for the $\rho_i^*$,

$$\int dx_1 \psi_j(x_1)\left[H^*(x_1,\cdots,x_n)\sum_{i=1}^{N}\psi_i^*(x_1)\rho_i^*(x_2,\cdots,x_n) - f^*(x_1,\cdots,x_n)\right] = 0, \qquad j = 1,\cdots,N. \tag{2.34}$$

A comparison of Equations 2.3 and 2.33 reveals that the choice

$$\alpha_j(x_1,\cdots,x_n) = \psi_j^*(x_1), \tag{2.35}$$

leads to a set of approximate equations which have the property that their solution, when used in Equations 2.32, renders the variational functional stationary; that is,

$$L(\phi^*,\phi) = \int dx_n \cdots \int dx_1 f^*(x_1,\cdots,x_n)\phi_0(x_1,\cdots,x_n) + \text{order}\,(\delta\phi^*,\delta\phi). \tag{2.36}$$

In many systems, the stationary value of the functional, described by the first term in Equation 2.36, has a physical interpretation.

When $H$ is a self-adjoint, positive definite operator; that is,

$$\int dx_n \cdots \int dx_1 u H\phi = \int dx_n \cdots \int dx_1 \phi H u,$$

and

$$\int dx_n \cdots \int dx_1 \phi H \phi \geqslant 0,$$

the Lagrangian functional becomes a minimum variational principle,

$$L(\phi^*, \phi) = \int dx_n \cdots \int dx_1 [2f\phi - \phi H \phi], \tag{2.37}$$

for which the theory has been extensively developed.[43]

A variational principle for the eigenvalue problem,

$$[H(x_1, \cdots, x_n) - \lambda K(x_1, \cdots, x_n)]\phi(x_1, \cdots, x_n) = 0, \tag{2.38}$$

is given by the functional,[48]

$$\lambda(\phi^*, \phi) = \frac{\int dx_n \cdots \int dx_1 \phi^*(x_1, \cdots, x_n) H(x_1, \cdots, x_n) \phi(x_1, \cdots, x_n)}{\int dx_n \cdots \int dx_1 \phi^*(x_1, \cdots, x_n) K(x_1, \cdots, x_n) \phi(x_1, \cdots, x_n)}, \tag{2.39}$$

where $\phi^*$ satisfies the adjoint equation

$$[H^*(x_1, \cdots, x_n) - \lambda K^*(x_1, \cdots, x_n)]\phi^*(x_1, \cdots, x_n) = 0. \tag{2.40}$$

When expansions of the form of Equations 2.32 are substituted into Equation 2.39 and the functional is required to be stationary with respect to arbitrary variations of $\rho_i$ and $\rho_i^*$, $N$ equations for $\rho_i$,

$$\int dx_1 \psi_j^*(x_1)[H(x_1, \cdots, x_n) - \lambda K(x_1, \cdots, x_n)] \sum_{i=1}^{N} \psi_i(x_1)\rho_i(x_2, \cdots, x_n) = 0, \qquad j = 1, \cdots, N, \tag{2.41}$$

and $N$ equations for $\rho_i^*$,

$$\int dx_1 \psi_j(x_1)[H^*(x_1, \cdots, x_n) - \lambda K^*(x_1, \cdots, x_n)] \sum_{i=1}^{N} \psi_i^*(x_1)\rho_i^*(x_2, \cdots, x_n) = 0, \qquad j = 1, \cdots, N, \tag{2.42}$$

are obtained. The bilinear concomitant associated with $H$ was assumed to vanish in writing Equations 2.42.

Comparison of Equations 2.41 and 2.3 indicates that using

$$\alpha_j(x_1, \cdots, x_n) = \psi_j^*(x_1), \tag{2.43}$$

in the Equations 2.3 leads to a determination of the $\rho_i$ such that solutions of the form of Equations 2.32 make the functional stationary.

The functional is stationary about some eigenvalue of Equation 2.38, but not necessarily the lowest one, unless $H$ is self-adjoint and positive definite. In this latter case, Equation 2.39 is a minimum principle for the lowest eigenvalue of the system.[44]

(*j*) *Emphasize certain intervals of* $x_1$. Problems may be encountered in which a relatively accurate solution is required only over a limited interval of $x_1$, or it may be desired to emphasize those intervals of $x_1$ in which a given physical process takes place. If there is sufficient physical insight, the weighting functions can be chosen to emphasize those intervals of $x_1$ that are most significant in determining a given aspect of of the solution.

While Condition 3, enumerated at the beginning of this section, serves as a general criterion for the selection of modes, a variety of criteria may be formulated for the weighting operators. Certain error criteria were discussed. Of these, the "minimal error criteria" of Equations 2.9 and 2.37, and Equation 2.39 when the operator is self-adjoint and positive definite, offer a certain advantage with respect to evaluating different sets of modes. If solutions are constructed according to these criteria for different sets of modes, a "best" choice of modes is that set which results in the least value of the functional.

Other criteria enter the consideration of a practical problem, however. The computational effort involved in using a weighting operator is usually much greater than when a simple weighting function is used. The simplest weighting functions are of types (*c*) and (*d*) mentioned earlier, and the use of the modes themselves as weighting functions is only slightly more complicated. Another advantage of using weighting functions rather than operators is that Equation 2.4 then has the same functional dependence upon $(x_2, \cdots, x_n)$ as would Equation 2.1 if the $x_1$ dependence was suppressed, and, consequently, solution techniques applicable to the latter may be adapted to the former.

## 2.3 Applications in Reactor Physics

Systematic application of the modal expansion method has been made to several problems in reactor physics. The majority of these applications are based on the variational principle of Equation 2.30, or a similar functional.

Modal expansion techniques were used to develop flux synthesis approximations for the spatial neutron distribution in three dimensions.[34,35,65] An expansion of the form

$$\phi(x, y, z) = \sum_{i=1}^{N} \psi_i(x, y)\rho_i(z)$$

was made and the appropriate equations of the form of Equations 2.3 were developed for the $\rho_i$. The modes $\psi_i$ can be obtained from two-dimensional calculations at various heights (values of $z$) in the reactor. The formalism was developed for four different types of weighting functions, which correspond to those of types (*c*), (*d*), and (*i*) of the previous section, and to the use of the modes as weighting functions.

Variational methods have been applied to obtain modal expansions for reactor-kinetics problems[19,34,36] and for depletion problems.[5] Expansions of the form

$$\phi(r, t) = \sum_{i=1}^{N} \psi_i(r)\rho_i(t),$$

were made with $\psi_i$ known, and Equations 2.3 were developed for the $\rho_i$ by using the procedures outlined in types (*i*)[19,36] and (*a*)[5] of the previous section. For the kinetics problems, the modes were specified either as the normal modes of the unperturbed problem, a combination of the normal modes of the perturbed and unperturbed problems, or as "Green's function modes."[19] In the depletion study, the modes were spatial distributions calculated at different times in life.

Modal expansion techniques have been used to calculate the spatially dependent thermal spectrum in thermal reactors. A variational technique equivalent to that of type (*i*) of the previous section was used to develop equations of the form of Equations 2.3.[6,7] The flux was expanded in a product series

$$\phi(r, E) = \sum_{i=1}^{N} \psi_i(E)\rho_i(r).$$

The modes $\psi_i$ were infinite media spectra, characteristic of various media encountered in such problems. An alternate expansion scheme,[54,55] in which the spatial distribution in each region was assumed parabolic, was used to develop equations of the form of Equations 2.3 for the energy-dependent combining coefficients $\rho_i(E)$. This method was also extended to the development of improved finite difference equations. Several choices of weighting functions were considered, but a numerical study[54] led to the conclusion that the variational weighting functions of Equation 2.35 were best.

The least-squares variational technique (type *a* of the previous section) has been used to develop a simplified model for fast-reactor calculations.[50] This is a special case of the modal expansion method in which the flux was expanded

$$\phi(r, E) = \rho_1\psi_1(r, E) + \rho_2\psi_2(r, E),$$

where $\psi_1$ and $\psi_2$ are specified distributions. Equations 2.3 are algebraic in this case.

The previous examples are explicit applications of the modal expansion method, albeit most of them were derived from a variational principle without recourse to the more general formalism. Many standard approximations which are used to reduce the neutron distribution equation to a form which is more amenable to solution may be characterized as modal expansion techniques in the sense that they can be described by the procedure outlined in Section 2.1. Such a classification is useful in that it

1. Unifies many approximate techniques under a general formalism.
2. Provides a consistent basis of comparison among alternative approximations.
3. Contributes to an understanding of the type of approximation involved.
4. May suggest certain refinements or alternatives.

Either the spherical harmonics or the discrete ordinates approximation is commonly used to treat the angular dependence of the neutron distribution. In the former case, an explicit application of the modal expansion method is made. An expansion of the form of Equation 2.2 is performed, the spherical harmonics are used as modes and weighting functions, and Equations 2.3 become the familiar spherical harmonics form of the Boltzmann equation.[66] In the discrete ordinates approximation, the neutron distribution equation is written for several discrete values of the angular variable, and integrals over angle are replaced by quadrature formulas involving the neutron distribution function, evaluated at the several discrete values of the angular variable.[8] The functional dependence of the neutron distribution between the discrete values of the angular variable is assumed. This corresponds to a modal expansion in which mode $\psi_n$ has the assumed angular dependence in some interval about the $n$th discrete value of the angular variable, and vanishes outside this interval. Intervals corresponding to the different modes are complementary and nonoverlapping. The $n$th weighting function has a constant value within the interval about the $n$th discrete value of the angular variable, and vanishes outside this interval.

The standard technique of expanding the spatial dependence of the neutron distribution in eigenfunctions of the Helmholtz equation, in bare reactor problems, is an example of the use of the modal expansion method to treat the spatial variable. Another example is the commonplace use of a buckling term to represent the transverse leakage. This is equivalent to expanding the flux in a single transverse spatial mode, that characterized by the buckling.

In the multigroup approximation, a neutron balance equation is written for the integral of the neutron flux over each of several complementary and nonoverlapping energy intervals. Integrals over energy are approximated by quadratures involving these flux integrals. This corresponds to a modal expansion in which the $n$th mode has a specified distribution within the $n$th energy interval and vanishes outside this interval. These intervals are the group widths. The $n$th weighting function is unity within the $n$th group and zero elsewhere.

It is common practice in developing finite difference equations for the spatial distribution problem to integrate the neutron balance equation over a specified volume surrounding each node, resulting in a set of equations for the flux integrals over the volume about each node. These equations could also be derived by a modal expansion. The $n$th mode, representing the flux integral about the $n$th node, is a function which has a constant value over the volume about the $n$th node and vanishes outside this volume. The weighting functions are the same as the modes.

These examples illustrate the inclusive character of the modal expansion method. The characterization of the multigroup and finite difference approximations as modal expansions will be useful in qualitatively comparing these standard methods with the approximate methods developed in the next chapter.

# 3. Modal Approximations to the Stationary Neutron Distribution Equation

In this chapter, the modal expansion formalism of the previous chapter is systematically applied to obtain approximate equations for the neutron distribution function, and standard techniques are adapted to solve these equations. The choice of modes and weighting functions is discussed from a physical viewpoint. A comparison of the modal and conventional (multigroup, finite difference) approximations is made.

## 3.1 Modal Transport Approximation

The integrodifferential Boltzmann equation is usually taken as the starting point in rigorous analyses of the neutron distribution. For a stationary distribution, this may be written

$$H(\mathbf{r}, E, \boldsymbol{\Omega})\phi(\mathbf{r}, E, \boldsymbol{\Omega}) - f(\mathbf{r}, E, \boldsymbol{\Omega}) = 0, \tag{3.1}$$

where

$$\begin{aligned} H(\mathbf{r}, E, \boldsymbol{\Omega}) \equiv{}& \boldsymbol{\Omega} \cdot \nabla \_\_\_ + \Sigma_t(\mathbf{r}, E) \_\_\_ \\ & - \int_E^{\infty} dE' \int_{\Omega'} d\boldsymbol{\Omega}' \Sigma_{es}(\mathbf{r}, E'\boldsymbol{\Omega}'/E\boldsymbol{\Omega}) \_\_\_ - \int_E^{\infty} dE' \int_{\Omega'} d\boldsymbol{\Omega}' \Sigma_{in}(\mathbf{r}, E'\boldsymbol{\Omega}'/E\boldsymbol{\Omega}) \_\_\_ \\ & - \chi(E) \int_0^{\infty} dE' \int_{\Omega'} d\boldsymbol{\Omega}' [\nu\Sigma_f(\mathbf{r}, E')/4\pi] \_\_\_, \end{aligned} \tag{3.2}$$

where $\Sigma_{es}(\Sigma_{in}) \cdot dE\, d\boldsymbol{\Omega}$ is the probability that a neutron with coordinates

($\mathbf{r}$, $E'$, $\mathbf{\Omega}'$) has an elastic (inelastic) collision which leaves it with final coordinates within the differential element $dE\,d\mathbf{\Omega}$ about ($\mathbf{r}$, $E$, $\mathbf{\Omega}$). Also, $\Sigma_t$, $\nu\Sigma_f$, and $\chi$ are the total cross section, the fission cross section times the number of neutrons produced per fission, and the fission spectrum, respectively; $\mathbf{r}$, $E$, and $\mathbf{\Omega}$ refer to the spatial coordinate ($x$, $y$, $z$), the energy, and the direction of the neutron. Here $\phi$ and $f$ are the neutron flux and source, respectively, per unit energy, per unit volume, per unit solid angle.

Modal expansions could conceivably be performed with respect to the spatial, energy, and/or angular variable. Expansions with respect to the spatial and angular variables have been considered in some detail.[52,56] In this section, a modal expansion, in known functions of energy $\psi_i(E)$, is developed. The expansion

$$\phi(\mathbf{r}, E, \mathbf{\Omega}) = \sum_{i=1}^{N} \psi_i(E)\rho_i(\mathbf{r}, \mathbf{\Omega}) \tag{3.3}$$

may be substituted into Equation 3.1. The resulting equation is then multiplied by a weighting function $\alpha_j(E)$ and integrated over energy. When this is repeated for $N$ linearly independent weighting functions, and the following definitions are used,

$$\Sigma_t^{ji}(\mathbf{r}) \equiv \int_0^\infty dE\alpha_j(E)\Sigma_t(\mathbf{r}, E)\psi_i(E), \tag{3.4a}$$

$$\Sigma_s^{ji}(\mathbf{r}, \mathbf{\Omega}'/\mathbf{\Omega}) \equiv \int_0^\infty dE\alpha_j(E)\left[\int_E^\infty dE'\{\Sigma_{es}(\mathbf{r}, E'\mathbf{\Omega}'/E\mathbf{\Omega}) + \Sigma_{in}(\mathbf{r}, E'\mathbf{\Omega}'/E\mathbf{\Omega})\}\psi_i(E')\right], \tag{3.4b}$$

$$\nu\Sigma_f^{ji}(\mathbf{r}) \equiv \int_0^\infty dE\alpha_j(E)\chi(E)\int_0^\infty dE'\nu\Sigma_f(\mathbf{r}, E')\psi_i(E'), \tag{3.4c}$$

$$\Delta^{ji} \equiv \int_0^\infty dE\alpha_j(E)\psi_i(E), \tag{3.4d}$$

$$f^j(\mathbf{r}, \mathbf{\Omega}) \equiv \int_0^\infty dE\alpha_j(E)f(\mathbf{r}, E, \mathbf{\Omega}), \tag{3.4e}$$

a matrix equation, involving matrices of rank $N$, is obtained:

$$\Omega \cdot \nabla\mathsf{\rho}(\mathbf{r}, \mathbf{\Omega}) + \mathsf{\Sigma}_t(\mathbf{r}) \cdot \mathsf{\rho}(\mathbf{r}, \mathbf{\Omega}) - \int_{\Omega'} d\mathbf{\Omega}'\mathsf{\Sigma}_s(\mathbf{r}, \mathbf{\Omega}'/\mathbf{\Omega}) \cdot \mathsf{\rho}(\mathbf{r}, \mathbf{\Omega}') - \frac{1}{4\pi}\int_{\Omega'} d\mathbf{\Omega}'\nu\mathsf{\Sigma}_f(\mathbf{r}) \cdot \mathsf{\rho}(\mathbf{r}, \mathbf{\Omega}') - \mathsf{f}(\mathbf{r}, \mathbf{\Omega}) = 0. \tag{3.5}$$

The matrix elements are defined by Equations 3.4, except for

$$\Omega \equiv \begin{bmatrix} \Delta^{11}\boldsymbol{\Omega} & \cdots & \Delta^{1N}\boldsymbol{\Omega} \\ \vdots & \ddots & \vdots \\ \Delta^{N1}\boldsymbol{\Omega} & \cdots & \Delta^{NN}\boldsymbol{\Omega} \end{bmatrix},$$

and $\mathsf{p}$ and $\mathsf{f}$, which latter are just the column vectors formed by the $\rho_i$ and $f^i$. In the previous equations, boldface type refers to a vector in the ordinary sense, and a symbol in light sans serif type refers to a matrix (or column vector).

Appropriate boundary conditions for $\mathsf{p}$ may be derived from the exact interface and nonreentrant surface conditions.[14]

$$\phi(0+, E, \boldsymbol{\Omega}) = \phi(0-, E, \boldsymbol{\Omega}), \qquad \text{interface at } \mathbf{r} = 0,$$

$$\phi(\mathbf{a}, E, \boldsymbol{\Omega}) = 0, \quad \boldsymbol{\Omega} \cdot \mathbf{n} > 0, \qquad \text{nonreentrant surface at } \mathbf{r} = \mathbf{a}.$$

Substituting Equation 3.3 into these conditions

$$\sum_{i=1}^{N} \psi_i(E)[\rho_i(0+, \boldsymbol{\Omega}) - \rho_i(0-, \boldsymbol{\Omega})] = 0,$$

$$\sum_{i=1}^{N} \psi_i(E)[\rho_i(\mathbf{a}, \boldsymbol{\Omega})] = 0, \qquad \boldsymbol{\Omega} \cdot \mathbf{n} > 0,$$

and making use of the linear independence of the $\psi_i$ result in

$$\mathsf{p}(0+, \boldsymbol{\Omega}) = \mathsf{p}(0-, \boldsymbol{\Omega}), \qquad \text{interface,} \tag{3.6}$$

$$\mathsf{p}(\mathbf{a}, \boldsymbol{\Omega}) = 0, \quad \boldsymbol{\Omega} \cdot \mathbf{n} > 0, \qquad \text{nonreentrant surface.} \tag{3.7}$$

Equation 3.5 is formally identical to the multigroup transport equation, except that $\Omega$ is diagonal in the latter. This is not surprising because the multigroup approximation is a special case of a modal expansion, as was pointed out in Section 2.3. This analogy can be exploited to adapt solution techniques which have been developed for the multigroup transport equation[11,67] to the solution of Equation 3.5.

## 3.2 Modal Diffusion Approximation

The standard $P_1$ equations are obtained from Equation 3.1 by a modal expansion of the form

$$\phi(\mathbf{r}, E, \mu) = P_0(\mu)\phi_0(\mathbf{r}, E) + P_1(\mu)\phi_1(\mathbf{r}, E), \tag{3.8}$$

where $\mu$ is the cosine of the angle which the neutron direction vector $\boldsymbol{\Omega}$ makes with the polar axis, and the neutron distribution is assumed azimuthally symmetric about this axis; $P_0$ and $P_1$ are the Legendre polynomials of zero and first order. When Equation 3.8 is substituted into Equation 3.1, and the result is weighted in turn by $P_0$ and $P_1$ and integrated over angle, the two $P_1$ equations obtain (see Appendix G). These equations may be combined to yield a diffusion equation

$$H(\mathbf{r}, E)\phi_0(\mathbf{r}, E) - f(\mathbf{r}, E) = 0, \tag{3.9}$$

where

$$\begin{aligned} H(\mathbf{r}, E) \equiv & - \nabla \cdot [3\Sigma_{tr}(\mathbf{r}, E)]^{-1}\nabla \underline{\quad} + \Sigma_{tr}(\mathbf{r}, E) \underline{\quad} \\ & - \int_E^\infty dE'\Sigma_{et}(\mathbf{r}, E'/E) \underline{\quad} - \int_E^\infty dE'\Sigma_{in}(\mathbf{r}, E'/E) \underline{\quad} \\ & - \chi(E) \int_0^\infty dE'\nu\Sigma_f(\mathbf{r}, E') \underline{\quad}\,. \end{aligned} \tag{3.10}$$

The source and the inelastic scattering are assumed to be isotropic in the laboratory system, and the transport approximation (Appendix G) is used to account for the anisotropy in the elastic scattering.

In Equation 3.10,

$$\begin{aligned} \Sigma_{tr}(\mathbf{r}, E) &\equiv \Sigma_t(\mathbf{r}, E) - \Sigma_{es}^1(\mathbf{r}, E), \\ \Sigma_{et}(\mathbf{r}, E'/E) &\equiv \Sigma_{es}^0(\mathbf{r}, E'/E) - \Sigma_{es}^1(\mathbf{r}, E'/E), \\ \Sigma_{et}(\mathbf{r}, E) &\equiv \int_0^\infty dE'\Sigma_{et}(\mathbf{r}, E'/E), \end{aligned} \tag{3.11}$$

and

$$\begin{aligned} \Sigma_{es}^l(\mathbf{r}, E'/E) &\equiv \int_{\Omega_0} d\mathbf{\Omega}_0 P_l(\mu_0)\Sigma_{es}(\mathbf{r}, E'\mathbf{\Omega}'/E\mathbf{\Omega}), \\ \Sigma_{es}^l(\mathbf{r}, E') &\equiv \int_0^\infty dE\Sigma_{es}^l(\mathbf{r}, E'/E). \end{aligned} \tag{3.12}$$

In the first of Equations 3.12, the subscript 0 on $\mathbf{\Omega}$ and $\mu$ indicates the scattering angle, that is, the angle between the initial $\mathbf{\Omega}'$ and final $\mathbf{\Omega}$ directions.

These equations are frequently referred to as consistent $P_1$ theory, or diffusion theory with the transport approximation.[53] Other forms of diffusion theory have been derived from modal expansions differing from Equation 3.8.[56] Equation 3.9 is the starting point for the modal approximations to the space and energy dependence which follow.

3.2.1 *Direct Method.* The scalar flux (the zero subscript is dropped) is expanded in known functions of space and energy $\psi_i(\mathbf{r}, E)$,

$$\phi(\mathbf{r}, E) = \sum_{i=1}^{N} \rho_i \psi_i(\mathbf{r}, E). \quad (3.13)$$

This expansion is substituted into Equation 3.9, and the resulting equation is multiplied by a weighting operator $\alpha_j(\mathbf{r}, E)$ and integrated over energy and volume. When this is repeated for $N$ different weighting operators, $N$ coupled algebraic equations are obtained.

$$\int_0^\infty dE \int_V d\mathbf{r}\alpha_j(\mathbf{r}, E)\left[H(\mathbf{r}, E)\sum_{i=1}^{N} \rho_i\psi_i(\mathbf{r}, E) - f(\mathbf{r}, E)\right] = 0,$$
$$= 1,\cdots, N. \quad (3.14)$$

These may be written as a matrix equation

$$\mathsf{H}\cdot\mathsf{\rho} - \mathsf{f} = 0. \quad (3.15)$$

$\mathsf{H}$ is an N × N matrix, while $\mathsf{\rho}$ and $\mathsf{f}$ are column matrices of N elements each. The matrix elements are

$$H^{ji} \equiv \int_V d\mathbf{r} \int_0^\infty dE\alpha_j(\mathbf{r}, E)\Big[\Sigma_{tr}(\mathbf{r}, E)\psi_i(\mathbf{r}, E) - \nabla\cdot(3\Sigma_{tr}(\mathbf{r}, E))^{-1}\nabla\psi_i(\mathbf{r}, E)$$
$$- \int_E^\infty dE'\{\Sigma_{et}(\mathbf{r}, E'/E) + \Sigma_{in}(\mathbf{r}, E'/E)\}\psi_i(\mathbf{r}, E')$$
$$-\chi(E)\int_0^\infty dE'\nu\Sigma_f(\mathbf{r}, E')\psi_i(\mathbf{r}, E')\Big], \quad (3.16a)$$

$$f^j \equiv \int_V d\mathbf{r}\int_0^\infty dE\alpha_j(\mathbf{r}, E)f(\mathbf{r}, E). \quad (3.16b)$$

Equation 3.15 is an algebraic equation which may be solved by standard inversion techniques.

$$\mathsf{\rho} = \mathsf{H}^{-1}\cdot\mathsf{f}. \quad (3.17)$$

When there is no source, $\mathsf{f} \equiv 0$, the condition for Equation 3.15 to have a nontrivial solution is

$$det\,\mathsf{H} = 0. \quad (3.18)$$

Equation 3.18 is the criticality condition. In this case, the $\rho_i$ may be determined to within an arbitrary multiplicative constant, which may be determined from a normalization condition.

3.2.2 *Semidirect Method: Space Modes.* Spatial modes $\psi_i(\mathbf{r})$ are specified and the scalar flux is expanded,

$$\phi(\mathbf{r}, E) = \sum_{i=1}^{N} \rho_i(E)\psi_i(\mathbf{r}). \quad (3.19)$$

Because it is desired to exploit the formal analogy of the modal approximation with the conventional neutron balance equation for an infinite medium, the class of weighting operators considered will be limited to *functions* of the spatial variable (see Section 2.1 in this connection). Equation 3.19 may be substituted into Equation 3.9, multiplied by the weighting function $\alpha_j(\mathbf{r})$ and integrated over the volume of the reactor. When this is repeated for $N$ different weighting functions, $N$ coupled integral equations obtain,

$$\int_V d\mathbf{r}\alpha_j(\mathbf{r})\left[H(\mathbf{r}, E)\sum_{i=1}^{N}\psi_i(\mathbf{r})\rho_i(E) - f(\mathbf{r}, E)\right] = 0, \qquad j = 1,\cdots, N. \tag{3.20}$$

These $N$ equations may be written as a matrix equation

$$\Sigma(E)\cdot\rho(E) = \int_E^\infty dE'\Sigma_s(E'/E)\cdot\rho(E') + \chi(E) \int_0^\infty dE'\nu\Sigma_f(E')\cdot\rho(E') + \mathrm{f}(E), \tag{3.21}$$

when the matrix elements are defined as

$$\Sigma^{ji}(E) \equiv \int_V d\mathbf{r}\alpha_j(\mathbf{r})[\Sigma_{tr}(\mathbf{r}, E)\psi_i(\mathbf{r}) - \nabla\cdot(3\Sigma_{tr}(\mathbf{r}, E))^{-1}\nabla\psi_i(\mathbf{r})], \tag{3.22a}$$

$$\Sigma_s^{ji}(E'/E) \equiv \int_V d\mathbf{r}\alpha_j(\mathbf{r})[\Sigma_{et}(\mathbf{r}, E'/E) + \Sigma_{in}(\mathbf{r}, E'/E)]\psi_i(\mathbf{r}), \tag{3.22b}$$

$$\nu\Sigma_f^{ji}(E) \equiv \int_V d\mathbf{r}\alpha_j(\mathbf{r})\nu\Sigma_f(\mathbf{r}, E)\psi_i(\mathbf{r}), \tag{3.22c}$$

$$f^j(E) \equiv \int_V d\mathbf{r}\alpha_j(\mathbf{r})f(\mathbf{r}, E), \qquad i, j = 1,\cdots, N, \tag{3.22d}$$

and $\rho$ is the column matrix formed by the $\rho_i$.

Equation 3.21 is formally identical with the spatially independent neutron balance equation. The latter is a special case of the former, in which the spatial dependence is expanded in a single mode. Thus, techniques employed to solve the spatially independent neutron balance equation can be adapted to solve Equation 3.21. In particular, Equation 3.21 can be cast in multigroup format by integrating from $E_{n-1}$ to $E_n$.

$$\Sigma^n\cdot\rho^n = \sum_{n'=n}^{M}\Sigma_s^{n'/n}\cdot\rho^{n'} + \chi^n\sum_{n'=1}^{M}\nu\Sigma_f^{n'}\cdot\rho^{n'} + \mathrm{f}^n, \qquad n = 1,\cdots, M, \tag{3.23}$$

where $M$ is the number of groups. If the within group scattering is removed from both sides of Equations 3.23 by defining a removal matrix

$$\Sigma_R^n \equiv \Sigma^n - \Sigma_s^{n/n}, \tag{3.24}$$

equations result which are formally identical to the conventional multigroup equations.

$$\Sigma_R^n \cdot \rho^n = \sum_{n'=n+1}^{M} \Sigma_s^{n'/n} \cdot \rho^{n'} + \chi^n \sum_{n'=1}^{M} \nu\Sigma_f^{n'} \cdot \rho^{n'} + \mathrm{f}^n, \qquad n = 1,\cdots, M. \tag{3.25}$$

Standard multigroup definitions have been used in the foregoing equations; that is, the matrix elements of Equations 3.25 are

$$\rho_i^n \equiv \int_{E_{n-1}}^{E_n} dE\rho_i(E), \qquad f^{j,n} \equiv \int_{E_{n-1}}^{E_n} dEf^j(E),$$

$$\chi^n \equiv \int_{E_{n-1}}^{E_n} dE\chi(E),$$

$$\Sigma_s^{ji,n'/n} \equiv \int_{E_{n-1}}^{E_n} dE \int_{E_{n'-1}}^{E_{n'}} dE'\Sigma_s^{ji}(E'/E)\rho_i(E')/\rho_i^{n'}, \tag{3.26}$$

$$\nu\Sigma_f^{ji,n} \equiv \int_{E_{n-1}}^{E_n} dE\nu\Sigma_f^{ji}(E)\rho_i(E)/\rho_i^n,$$

$$\Sigma_R^{ji,n} \equiv \left[\int_{E_{n-1}}^{E_n} dE\Sigma^{ji}(E)\rho_i(E)/\rho_i^n\right] - \Sigma_s^{ji,n/n}.$$

These quantities, except $\rho_i^n$ and $f^{i,n}$, of course, can be evaluated from multigroup cross-section data.

Equations 3.25 may be written as a supermatrix equation

$$\begin{bmatrix} \Sigma_R^1 & 0 & \cdots & 0 \\ \vdots & & & \vdots \\ 0 & \Sigma_R^2 & \cdots & 0 \\ \vdots & & \ddots & \vdots \\ 0 & \cdots & & \Sigma_R^M \end{bmatrix} \begin{bmatrix} \rho^1 \\ \cdot \\ \cdot \\ \cdot \\ \rho^M \end{bmatrix} = \begin{bmatrix} 0 & 0 & \cdots & 0 & 0 \\ & & & & \\ \Sigma_s^{1/2} & 0 & \cdots & 0 & 0 \\ \vdots & & \ddots & & \vdots \\ \Sigma_s^{1/M} & \Sigma_s^{2/M} & \cdots & \Sigma_s^{M-1/M} & 0 \end{bmatrix} \begin{bmatrix} \rho^1 \\ \cdot \\ \cdot \\ \cdot \\ \rho^M \end{bmatrix}$$
$$+ \begin{bmatrix} \chi^1 \\ \\ \chi^2 \\ \vdots \\ \chi^M \end{bmatrix} [\nu\Sigma_f^1 \quad \cdots \quad \nu\Sigma_f^M] \begin{bmatrix} \rho^1 \\ \cdot \\ \cdot \\ \cdot \\ \rho^M \end{bmatrix} + \begin{bmatrix} \mathrm{f}^1 \\ \cdot \\ \cdot \\ \cdot \\ \mathrm{f}^M \end{bmatrix} \tag{3.27}$$

Equation 3.27 is formally identical to the conventional matrix form of the spatially independent multigroup equations. In the latter, the matrix elements are scalars, whereas in Equation 3.27, the matrix elements are $N \times N$ submatrices, $N$ being the number of expansion modes. The superscripts refer to the energy group. The solution procedure for Equation 3.27 is formally identical to the solution procedure for the spatially independent multigroup equations with $N \times M$ groups. Thus, the formal effect of the modal expansion is to expand the conventional multigroup matrix.

The significance of this result is that spatially dependent problems can be treated with techniques which are commonly employed for spatially independent problems (e.g., the spatially dependent spectrum in a multiregion reactor can be calculated with an infinite medium slowing down code). Spatial effects are incorporated into the calculation by the bilinear averaging of Equations 3.22. In particular, the leakage is accounted for by a "buckling" type term (the second term in Equation 3.22.*a*) for each mode.

3.2.3 *Semidirect Method: Energy Modes.* In this case, the scalar flux is expanded in known functions of energy $\psi_i(E)$,

$$\phi(\mathbf{r}, E) = \sum_{i=1}^{N} \psi_i(E)\rho_i(\mathbf{r}). \tag{3.28}$$

The class of weighting operators is limited to *functions* of energy, in order to exploit the formal analogy between the modal and multigroup approximations. Equation 3.28 is substituted into Equation 3.9, the resulting equation is multiplied by $\alpha_j(E)$, and integrated over energy. When this operation is repeated for $N$ different weighting functions, $N$ coupled partial differential equations are obtained.

$$\int_0^{\infty} dE\alpha_j(E)\left[H(\mathbf{r}, E)\sum_{i=1}^{N} \psi_i(E)\rho_i(\mathbf{r}) - f(\mathbf{r}, E)\right] = 0, \qquad j = 1, \cdots, N. \tag{3.29}$$

These equations may be written as a matrix equation,

$$-\nabla \cdot \mathsf{D}(\mathbf{r}) \cdot \nabla \mathsf{\rho}(\mathbf{r}) + \mathsf{\Sigma}(\mathbf{r}) \cdot \mathsf{\rho}(\mathbf{r}) - \mathsf{f}(\mathbf{r}) = 0, \tag{3.30}$$

where the matrix elements are defined as

$$D^{ji}(\mathbf{r}) \equiv \int_0^{\infty} dE\alpha_j(E)[3\Sigma_{tr}(\mathbf{r}, E)]^{-1}\psi_i(E), \tag{3.31a}$$

$$\Sigma^{ji}(\mathbf{r}) \equiv \int_0^\infty dE\alpha_j(E)\Big[\Sigma_{tr}(\mathbf{r}, E)\psi_i(E) - \int_E^\infty dE'\Sigma_{et}(\mathbf{r}, E'/E)\psi_i(E')$$

$$- \int_E^\infty dE'\Sigma_{in}(\mathbf{r}, E'/E)\psi_i(E')$$

$$-\chi(E)\int_0^\infty dE'\nu\Sigma_f(\mathbf{r}, E')\psi_i(E')\Big], \tag{3.31b}$$

$$f^j(\mathbf{r}) \equiv \int_0^\infty dE\alpha_j(E)f(\mathbf{r}, E), \tag{3.31c}$$

and $\boldsymbol{\rho}$ is the column matrix formed by the $\rho_i$.

Appropriate boundary conditions for Equation 3.30 can be derived from standard diffusion theory boundary conditions.[15] The continuity of flux and current across an interface at $\mathbf{r} = 0$ is expressed by

$$\phi(0+, E) = \phi(0-, E),$$

and

$$[3\Sigma_{tr}(0+, E)]^{-1}\nabla\phi(0+, E) = [3\Sigma_{tr}(0-, E)]^{-1}\nabla\phi(0-, E).$$

By substituting Equation 3.28, these become

$$\sum_{i=1}^N \psi_i(E)[\rho_i(0+) - \rho_i(0-)] = 0,$$

and

$$\sum_{i=1}^N \psi_i(E)\{[3\Sigma_{tr}(0+, E)]^{-1}\nabla\rho_i(0+) - [3\Sigma_{tr}(0-, E)]^{-1}\nabla\rho_i(0-)\} = 0.$$

Because the $\psi_i(E)$ are linearly independent, the first of these equations requires that each $\rho_i$ be continuous across an interface. To obtain $N$ conditions on the $\rho_i$ from the second of these equations, it is multiplied, in turn, by each of the weighting functions and integrated over energy. Thus, the scalar flux is continuous across the interface at all energies, and the current is continuous in a weighted integral sense. The resulting boundary conditions are

$$\begin{aligned} \boldsymbol{\rho}(0+) &= \boldsymbol{\rho}(0-), \\ \mathbf{D}(0+)\cdot\nabla\boldsymbol{\rho}(0+) &= \mathbf{D}(0-)\cdot\nabla\boldsymbol{\rho}(0-), \quad \text{interface.} \end{aligned} \tag{3.32}$$

The vanishing of the flux at an extrapolated surface $\mathbf{a}$,

$$\phi(\mathbf{a}, E) = 0,$$

is a special case of the first of Equations 3.32.

$$\boldsymbol{\rho}(\mathbf{a}) = 0, \quad \text{extrapolated nonreentrant surface.} \tag{3.33}$$

The zero-current condition is a special case of the second of Equations 3.32.

$$\mathsf{D}(0) \cdot \nabla \rho(0) = 0, \qquad \text{zero current.} \tag{3.34}$$

Equation 3.30 is formally identical to the matrix form of the multigroup diffusion equation, except that the matrix $\mathsf{D}$ is diagonal in the latter. Techniques used to solve the multigroup diffusion equation can, consequently, be exploited to solve Equation 3.30. Analytical solutions of this equation are developed, for simple geometries, in Appendix B. In Appendix C, a finite difference approximation to Equation 3.30 is developed for one-dimensional geometries.

## 3.3 Modes and Weighting Functions

General considerations relevant to the selection of modes and weighting operators were discussed in Section 2.2. It is the purpose of this section to motivate supplemental physical considerations that may be invoked in applying the formalism of the previous section. Three general criteria, which may be concluded from Chapter 2, form the basis of the discussion of this section.

1. The modes should be so chosen that the flux and the source are representable by a linear combination of the modes.
2. The modal approximation, Equations 2.3, is obtained by requiring that the modal expansion satisfies the governing equation in a weighted integral sense. Thus, the weighting function should describe the relative importance of the neutron balance equation being satisfied.
3. The modes must be linearly independent, and the weighting functions must be linearly independent.

3.3.1 *Modes.* An attempt to catalogue the means by which modes may be motivated and constructed would be premature prior to a widespread application of the modal expansion method. The intent of this section is rather to develop certain physical concepts which may motivate the choice of modes and to discuss several means of constructing modes.

(*a*) *Multiple source concept.* When it is meaningful to visualize the neutron distribution as the superposition of distributions emanating from different "sources," then the latter distributions may be feasible modes for the problem. For example, the infinite medium spectrum in a fast-reactor assembly consists of both neutrons elastically moderating from the fission spectrum, and also of neutrons which have had one or more inelastic collisions and are elastically moderating from the inelastic spectrum.

(*b*) *Coupled region concept.* A reactor core usually can be conceptually sectioned into several regions, within each of which there is no large-scale nonuniformity. If one of these regions is so large that the neutron distribution within it depends only slightly upon adjacent regions, then this region is considered to be *loosely coupled.* If, on the other hand, the neutron distribution within a region is highly sensitive to the composition of adjacent regions, this region is considered to be *tightly coupled.* Note that a region may be tightly or loosely coupled with respect to the spectrum, spatial distribution, or both.

When a region is loosely coupled, the stationary neutron distribution which would obtain in that region if it were isolated may be a feasible mode. For the modular core fast-reactor concept, where a number of cores are arrayed in a large block of blanket material, the distributions obtained when each of the cores, in turn, is assumed to be the only core in the array might be suitable space-energy modes for the direct method. Alternatively, if the cores are many mean free paths in dimension, the infinite media spectra of the various cores and a spectrum of neutrons moderating in the blanket material may be feasible energy modes for the semidirect method. The energy modes would be much easier to construct, in this case.

In tightly coupled regions, it is perhaps more appropriate to base a selection of modes on the extreme distributions which could obtain if certain regions or materials were removed. Energy modes for a tightly coupled multiregion fast-reactor assembly might be chosen in the following three ways:

1. The spectrum of neutrons elastically moderating from the fission spectrum according to the properties of the least effective moderating region.
2. The spectrum of neutrons elastically moderating from the inelastic spectrum according to the properties of the most effective moderating region.
3. The spectrum of neutrons elastically moderating in the presence of those resonance-scattering nuclei which are apt to have a significant effect upon the fine structure of the spectrum.

Modes 1 and 2 should be "harder" and "softer," respectively, than the actual spectrum, while mode 3 should reflect the fine structure due to resonances.

(*c*) *Experimental data.* Experimental data may be used to motivate and construct modes. Spatial traverses performed with threshold detectors (e.g., $\sigma_f^{28}$, $\sigma_f^{40}$) indicate the distribution of the neutron flux above threshold, in a gross manner. Similarly, resonance-detector traverses provide an indication of the spatial distribution of the flux about the resonance energy, and thermal-detector traverses indicate the spatial distribution of the thermal neutron flux. The $1/v$ detectors (e.g., $B^{10}$) yield information about the spatial distribution of the neutron number density. Several detectors with different thresholds or recoil-proton detectors may be employed to construct an approximate spectral distribution.

(*d*) *Other.* Intuition, experience with similar systems, and calculations based on simpler models may also be utilized to construct modes. For example, the spatial distributions of the group fluxes in a few group calculation may afford suitable spatial modes for a semidirect modal expansion.

3.3.2 *Weighting Functions.* The modal approximation, Equations 2.3, is obtained by requiring that the modal expansion satisfies the neutron balance equation in a weighted integral sense, for $N$ different weighting functions. These weighting functions should thus be chosen to describe the relative importance, in some sense, that the neutron balance equation be satisfied. The purposes of this section are to consider different senses in which it is important that the neutron balance equation is satisfied, to discuss the appropriate weighting functions, and to indicate means by which these functions can be constructed.

In static reactor calculations, many important parameters derive from integral quantities such as reaction rates and reactivity worths. Power distributions, reactivity coefficients, conversion ratio, critical mass, control rod worths, isotopic depletion rates, and so on, are examples of parameters, derived from integral quantities, which characterize the performance of a reactor.

From the discussion of type *b* of Section 2.2, it follows that a reaction rate

$$R = \int_0^\infty dE \int_V d\mathbf{r}\Sigma_a(\mathbf{r}, E)\phi(\mathbf{r}, E),$$

will be correctly predicted by the modal approximation if a weighting function is chosen as

$$\alpha(\mathbf{r}, E) = \phi^*(\mathbf{r}, E),$$

where $\phi^*$ is the importance function (adjoint) that satisfies

$$H^*(\mathbf{r}, E)\phi^*(\mathbf{r}, E) - \Sigma_a(\mathbf{r}, E) = 0. \tag{3.35}$$

The operator $H^*$ is mathematically adjoint to $H$ of Equation 3.10; that is,

$$H^*(\mathbf{r}, E) \equiv -\nabla \cdot [3\Sigma_{tr}(\mathbf{r}, E)]^{-1} \nabla \_\_\_ + \Sigma_{tr}(\mathbf{r}, E) \_\_\_$$
$$- \int_0^E dE' \Sigma_{et}(\mathbf{r}, E/E') \_\_\_ - \int_0^E dE' \Sigma_{in}(\mathbf{r}, E/E') \_\_\_$$
$$- \nu\Sigma_f(\mathbf{r}, E) \int_0^\infty dE' \chi(E') \_\_\_. \tag{3.36}$$

If a local reaction rate at $\mathbf{r}_0$ is required, the inhomogeneous term in Equation 3.35 should be

$$\Sigma_a(\mathbf{r}_0, E) \cdot \delta(\mathbf{r} - \mathbf{r}_0).$$

For a subcritical-reactor problem, importance functions proportional to the solution of Equation 3.35 for various $\Sigma_a$ can be used as weighting functions. As the reactor approaches critical, all of these solutions will coalesce into the solution of Equation 3.37.[58]

In a critical reactor, the importance function (adjoint) $\phi^*(\mathbf{r}, E)$, which satisfies

$$H^*(\mathbf{r}, E)\phi^*(\mathbf{r}, E) = 0, \tag{3.37}$$

is proportional to the asymptotic power level which results when a neutron is introduced at point $\mathbf{r}$ and energy $E$, and the reactor is initially at zero power. This interpretation may be induced from the concept of an iterated fission probability.[30] (See also Reference 37.) Because the amount by which the modal expansion fails to satisfy the neutron balance equation may be considered as a pseudo source, weighting with $\phi^*$ of Equation 3.37 ensures that the modal approximation will predict the correct net "worth" of the neutron distribution. Thus, the criticality of the reactor should be correctly predicted when the weighting function is chosen as the solution of Equation 3.37.

Obtaining the solution of the adjoint equation is at least as difficult as solving the flux equation. Hence, the previous remarks serve more to indicate the nature of the weighting functions than to provide a means for their construction.

In a reactor composed of several loosely coupled regions, the infinite medium importance functions for the several regions may serve as weighting functions, in those regions. Approximate calculations, based on crude models, and experience or intuition, may also serve to suggest appropriate weighting functions.

The importance function of Equation 3.35, in principle, may be determined from experiment.[58] To see this, multiply Equation 3.9 by

$\phi^*$ and Equation 3.35 by $\phi$, subtract the two, and integrate over volume and energy.

$$\int_0^\infty dE \int_V d\mathbf{r}\phi^*(\mathbf{r}, E)f(\mathbf{r}, E) = \int_0^\infty dE \int_V d\mathbf{r}\phi(\mathbf{r}, E)\Sigma_a(\mathbf{r}, E).$$

The right side can be measured. If a point, monoenergetic source is employed, that is, if

$$f(\mathbf{r}, E) = \delta(\mathbf{r} - \mathbf{r}_0) \cdot \delta(E - E_0),$$

then

$$\phi^*(\mathbf{r}_0, E_0) = \int_0^\infty dE \int_V d\mathbf{r}\phi(\mathbf{r}, E)\Sigma_a(\mathbf{r}, E). \tag{3.38}$$

When there is inadequate insight to enable the construction of importance functions, or when the computational effort which would be involved is deemed excessive, the modes themselves may be used as weighting functions. This may be crudely rationalized by the fact that such a choice emphasizes those regions in which the solution is largest.

## 3.4 Qualitative Comparison of Modal and Conventional Approximations

In this section, the energy modal (Section 3.2.3) and the space modal (Section 3.2.2) approximations are compared with conventional approximations of the energy and space dependence of the neutron flux. These latter are the multigroup approximation of the energy dependence and the finite difference approximation of the spatial dependence.

The multigroup approximation is a special case of the general energy modal approximation, in which the *n*th multigroup mode has a constant value over the width of the *n*th group and vanishes elsewhere. Computational effort required in solving an $N$-mode set of equations of the form of Equation 3.30 is roughly equal to that required in solving the $N$-group diffusion equations. Consequently, a comparison of the two methods reduces to a comparison of unrestricted and multigroup modes.

Multigroup modes are defined by specifying the group structure. Multigroup cross sections are then defined by assuming some within-group weighting spectrum and performing an average of the nuclear data.[68] Any physical insight into the nature of the solution may be incorporated into the weighting spectrum, and thus biases the solution with respect to the information incorporated. Representation of a rapidly varying spectrum requires the use of a large number of small width groups.

Unrestricted modes are considerably more flexible. Physical insight into the nature of the solution in some part of the reactor may be used to construct a single mode, but does not have to influence the other modes, and thus does not necessarily bias the solution in other parts of the reactor. Fine structure in the spectrum, due to resonances, may be incorporated into an energy mode, introducing the possibility that even a rapidly varying distribution can be represented by relatively few modes.

Measured reaction rates and reactivity worths can be incorporated into the energy modal approximation. The perturbation theory expression for the reactivity worth of a single atom in an essentially infinite medium, in the $P_1$ transport approximation, is

$$\delta k \sim \int_0^\infty dE\phi^*(E)\left[\sigma_{tr}(E)\phi(E) - \int_E^\infty dE'\sigma_s(E'/E)\phi(E') - \chi(E)\int_0^\infty dE'\nu\sigma_f(E')\phi(E')\right].$$

If mode $(i)$ and weighting function $(j)$ are chosen as the spectrum and importance function, respectively, in the region where the reactivity worth is measured, then $\Sigma^{ji}$ of Equation 3.31*b* can be constructed from measured reactivity worths. Conversely, parameters in the modal model may be unambiguously correlated with measured reactivity worths.

The reaction rate at any point in the reactor is

$$R(\mathbf{r}) = \int_0^\infty dE\Sigma_a(\mathbf{r}, E)\phi(\mathbf{r}\ E) \Rightarrow \sum_{i=1}^N \rho_i(\mathbf{r}) \int_0^\infty dE\Sigma_a(\mathbf{r}, E)\psi_i(E).$$

If mode $\psi_i$ is chosen as the spectrum which obtains at the point where reaction rates are measured, then the integral involving $\psi_i$ in the preceding expression can be constructed from measured reaction rates. This is in contrast to the multigroup approximation, in which it is not possible to incorporate integral measurements in a direct and unambiguous manner.

A small number of appropriately chosen energy modes may represent the spectrum throughout a simple reactor quite well. This suggests that the energy modal approximation may be used to generate a simple calculational model. This conjecture is substantiated in Chapter 5.

The finite difference approximation may be considered a special case of the general space modal approximation, in which the $n$th mode has a constant value over the volume associated with the $n$th mesh point and vanishes elsewhere. Solution of the $N$-mode approximation and the finite difference approximation with $N$ mesh points require roughly equal computational effort. Again, a comparison of the two approximations reduces to a comparison of the unrestricted and finite difference modes.

The finite difference modes are fixed by specifying the mesh network and the manner in which the flux is assumed to vary between mesh points. Physical insight as to the spatial variation of the flux at some energy may be incorporated into these modes via the assumed flux variation between mesh points. However, this will bias the equations at all energies. To represent a rapidly varying spatial distribution, it is necessary to employ a large number of closely spaced mesh points.

Physical insight concerning the spatial distribution of the flux at some energy may be incorporated into one of the unrestricted modes, without necessarily biasing the flux shape at other energies. Rapid variations in the spatial distribution may be reflected in one or more of the unrestricted modes, thus enabling a few suitably chosen modes to represent rapidly varying distributions.

The multigroup and finite difference equations do have the conceptual advantage of being physically interpretable as balance equations on a finite region of energy and space, respectively. No such appealing interpretation of the individual modal equations is possible, in general.

# 4. Spatially Independent Spectrum and Importance Function in Fast Reactors

From the discussion of the preceding chapter, it is apparent that spectra and importance functions associated with various media may serve as modes and weighting functions in the energy modal approximation. Although these functions could be obtained from a multigroup calculation, such a procedure would vitiate the appeal of the modal approximation as a means of obtaining relatively accurate results from a *simple* calculation. Thus, there is a need for simple means to construct spatially independent spectra and importance functions.

The purpose of this chapter is to develop analytical expressions for the spatially independent spectrum and importance function. This is accomplished by solving the neutron balance equation, and the equation adjoint thereto, by the method of successive approximations. In order to obtain the solutions in more tractable form, it is necessary to introduce certain approximations that render the solutions more suitable for media composed of intermediate and heavy atomic mass nuclei than for low A nuclei.

The validity of the analytical expressions is verified by comparison with experiment and multigroup calculations. The expressions are found to be satisfactory for both highly enriched and dilute fast-reactor assemblies.

## 4.1 Spectrum

The collision density is a smoother function of energy than the neutron flux in fast-reactor assemblies,[17] due principally to elastic scattering resonances in intermediate $A$ nuclei present in the form of structure and coolant. In this section, the neutron balance equation is written and solved in terms of the collision density, which, in the $P_1$ transport approximation (Appendix G), is the product of the neutron flux and the sum of the transport and leakage cross sections. For a critical medium containing $M$ different inelastic scattering species, the neutron balance equation is

$$F(E) = \chi(E) + \int_E^{E/\alpha} \frac{r_{et}(E')F(E')}{E'(1-\alpha)}\, dE' + \sum_{m=1}^{M} P^m(E) \int_0^\infty r_{in}^m(E')F(E')dE', \tag{4.1}$$

subject to the normalization

$$\int_0^\infty dE\nu r_f(E)F(E) = 1. \tag{4.2}$$

Here $F$, $\chi$, and $P^m$ are the collision density, fission spectrum, and inelastic spectrum of neutrons scattered from nuclear species $m$, respectively, per unit energy. The $r_{et}$, $r_f$, and $r_{in}^m$ are the elastic scattering probability, the fission probability, and the inelastic scattering probability for nuclear species $m$, respectively;

$$r_{et} \equiv \frac{\Sigma_{et}}{\Sigma_{tr} + \Sigma_L}, \qquad r_{in}^m \equiv \frac{\Sigma_{in}^m}{\Sigma_{tr} + \Sigma_L}, \qquad r_f \equiv \frac{\Sigma_f}{\Sigma_{tr} + \Sigma_L},$$

where $\Sigma_{tr}$ and $\Sigma_{et}$ are the transport and $P_1$ "transport" elastic scattering cross section, respectively, and are defined by Equations 3.11. The $\Sigma_{in}^m$, $\Sigma_f$, and $\Sigma_L$ are the inelastic scattering cross section for species $m$, the fission cross section, and the leakage cross section, respectively.

The fission spectrum for U-235 and Pu-239 is represented by,[63]

$$\chi(E) = \frac{2}{\sqrt{\pi}} \cdot \frac{E^{1/2}}{\theta^{3/2}} e^{-E/\theta}, \qquad \theta = 1.32 \text{ Mev}. \tag{4.3}$$

The inelastic spectrum for nuclear species $m$ may be represented by the evaporation model (Appendix D)

$$P^m(E) = \frac{Ee^{-E/T_m}}{T_m^{\ 2}}, \tag{4.4}$$

where $T_m$ is a parameter characterizing the inelastic spectrum of nuclear species $m$, and is assumed independent of incident neutron energy. Evaluation of $T_m$ is discussed in Appendix D.

Equation 4.1 may be solved by the method of successive approximations,[45] the algorithm for which is

$$F_0(E) = \chi(E), \tag{4.5}$$

$$F_n(E) = \chi(E) + \int_E^{E/\alpha} dE' \frac{r_{et}(E')F_{n-1}(E')}{E'(1-\alpha)}$$

$$+ \sum_{m=1}^{M} P^m(E) \int_0^\infty dE' r_{in}^m(E') F_{n-1}(E'), \; n \geqslant 1.$$

The solution (infinite-order approximation) obtained in this manner is

$$F(E) = \left[\chi(E) + \sum_{n=1}^{\infty} \bar{K}_n(E)\right] + \sum_{m=1}^{M} \beta^m \left[P^m(E) + \sum_{n=1} \bar{J}_n{}^m(E)\right], \tag{4.6}$$

where the iterated kernels, $\bar{K}_n$ and $\bar{J}_n{}^m$, are given by

$$\bar{K}_0(E) \equiv \chi(E),$$

$$\bar{K}_n(E) \equiv \int_E^{E/\alpha} \frac{r_{et}(E')\bar{K}_{n-1}(E')}{E'(1-\alpha)} dE', \qquad n \geqslant 1, \tag{4.7}$$

and

$$\bar{J}_0{}^m(E) \equiv P^m(E),$$

$$\bar{J}_n{}^m(E) \equiv \int_E^{E/\alpha} \frac{r_{et}(E')\bar{J}_{n-1}^m(E')}{E'(1-\alpha)} dE', \qquad n \geqslant 1. \tag{4.8}$$

The quantity $\beta^m$ is a constant defined by

$$\beta^m \equiv r^m + \sum_{i=1}^{M} r^i l^{mi} + \sum_{j=1}^{M} \sum_{i=1}^{M} r^i l^{ji} l^{mj} + \sum_{k=1}^{M} \sum_{j=1}^{M} \sum_{i=1}^{M} r^i l^{ji} l^{kj} l^{mk} + \cdots, \tag{4.9}$$

where

$$r^m \equiv \int_0^\infty dE r_{in}^m(E) \left[\sum_{n=1}^{\infty} \bar{K}_{n-1}(E)\right],$$

$$l^{mk} \equiv \int_0^\infty dE r_{in}^m(E) \left[\sum_{n=1}^{\infty} \bar{J}_{n-1}^k(E)\right]. \tag{4.10}$$

In order to obtain a closed-form solution, effective elastic scattering probabilities $\bar{r}_{et}$ and $\bar{r}_{et}^m$ are defined which result in modified iterated kernels.

$$\bar{K}_n(E) \Rightarrow (\bar{r}_{et}(E))^n \cdot K_n(E), \qquad \text{all } n \geqslant 1, \tag{4.11}$$

$$\bar{J}_n{}^m(E) \Rightarrow (\bar{r}_{et}^m(E))^n \cdot J_n(E), \qquad \text{all } n \geqslant 1, \tag{4.12}$$

where

$$K_0(E) \equiv \chi(E),$$
$$K_n(E) \equiv \int_E^{E/\alpha} \frac{K_{n-1}(E')dE'}{E'(1-\alpha)}, \qquad n \geqslant 1, \tag{4.13}$$

and

$$J_0{}^m(E) \equiv P^m(E),$$
$$J_n{}^m(E) \equiv \int_E^{E/\alpha} \frac{J_{n-1}^m(E')dE'}{E'(1-\alpha)}, \qquad n \geqslant 1. \tag{4.14}$$

Using Equations 4.3 and 4.4, these modified iterated kernels $K_n$ and $J_n{}^m$ may be evaluated analytically by making a series expansion in $(1-\alpha)$ and retaining only terms through first order. The expressions thus obtained lead to summable series when substituted into Equation 4.6, and the final expression for the collision density is

$$F(E) = K(E) + \sum_{m=1}^{M} \beta^m J^m(E), \tag{4.15}$$

where

$$K(E) \equiv \sum_{n=0}^{\infty} \bar{K}_n(E)$$
$$\Rightarrow \chi(E)\left[\frac{1}{1-\bar{r}_{et}(E)} + \frac{(1-\alpha)}{2}\left\{\frac{\bar{r}_{et}(E)}{(1-\bar{r}_{et}(E))^2}\left(\frac{3}{2}-\frac{E}{\theta}\right)\right\} + \cdots O((1-\alpha)^2)\right], \tag{4.16}$$

$$J^m(E) \equiv \sum_{n=0}^{\infty} \bar{J}_n{}^m(E)$$
$$\Rightarrow P^m(E)\left[\frac{1}{1-\bar{r}_{et}^m(E)} + \frac{(1-\alpha)}{2}\left\{\frac{\bar{r}_{et}^m(E)}{(1-\bar{r}_{et}^m(E))^2}\left(2-\frac{E}{T_m}\right)\right\} + \cdots O((1-\alpha)^2)\right], \tag{4.17}$$

and the constants involved in the definition of $\beta^m$, Equation 4.9, become

$$r^m \Rightarrow \int_0^\infty dE r_{in}^m(E) K(E), \qquad l^{mk} \Rightarrow \int_0^\infty dE r_{in}^m(E) J^k(E). \tag{4.18}$$

Once effective elastic scattering probabilities are defined which satisfy Relations 4.11 and 4.12, in some sense, the spectrum may be obtained from Equation 4.15 simply by evaluating the integrals of Equations 4.18 and performing a few multiplications. These effective elastic scattering probabilities may be motivated by a physical argument, and given mathematical form by employing continuous slowing down theory. This is discussed in Appendix E, where Equation 4.6 is derived from physical considerations without recourse to the neutron balance equation, and where the mathematical details of the foregoing derivation are also given.

## 4.2 Importance Function

The importance function for a critical spatially independent medium is defined by the equation mathematically adjoint[49] to Equation 4.1,

$$\phi^*(E) = \nu r_f(E) + \frac{r_{et}(E)}{E(1-\alpha)} \int_{\alpha E}^{E} \phi^*(E')dE' + \sum_{m=1}^{M} r_{in}^m(E) \int_0^\infty dE' P^m(E')\phi^*(E'), \quad (4.19)$$

where the normalization condition is

$$\int_0^\infty dE\chi(E)\phi^*(E) = 1. \quad (4.20)$$

A physical interpretation of the importance function as an iterated fission probability[30] was mentioned in Chapter 3. Equation 4.19 has the physical significance of being a statement of the property of conservation of importance, which must obtain in a critical system. Each term on the right side is a product of two factors: (1) the probability that an event (fission, scattering) takes place at energy $E$ which produces a given number of neutrons ($\nu$ or one), and (2) the importance of these neutrons weighted over the spectrum associated with the event (fission, scattering) which produced them. Note that a leakage or capture event produces zero neutrons, and thus does not appear directly in the equation.

Equation 4.19 can be solved by the method of successive approximations according to the algorithm

$$\begin{aligned} \phi_0^*(E) &= \nu r_f(E), \\ \phi_n^*(E) &= \nu r_f(E) + \frac{r_{et}(E)}{E(1-\alpha)} \int_{\alpha E}^{E} dE' \phi_{n-1}^*(E') \\ &\quad + \sum_{m=1}^{M} r_{in}^m(E) \int_0^\infty dE' P^m(E') \phi_{n-1}^*(E'), \qquad n \geqslant 1. \end{aligned} \quad (4.21)$$

The solution obtained in this manner is

$$\phi^*(E) = \left[\sum_{n=1}^{\infty} \bar{k}_n(E)\right] + \sum_{m=1}^{M} \left(\sum_{n=1}^{\infty} f_n{}^m\right)\left[\sum_{n=1}^{\infty} \bar{h}_n{}^m(E)\right.$$
$$+ \sum_{m'=1}^{M} \left(\sum_{n=1}^{\infty} \bar{h}_n{}^{m'}(E)\right)\left(\sum_{n=1}^{\infty} g_n{}^{m'm}\right)$$
$$\left.+ \sum_{m''=1}^{M} \sum_{m'=1}^{M} \left(\sum_{n=1}^{\infty} \bar{h}_n{}^{m'}(E)\right)\left(\sum_{n=1}^{\infty} g_n{}^{m'm''}\right)\left(\sum_{n=1}^{\infty} g_n{}^{m''m}\right) + \cdots\right]. \tag{4.22}$$

The iterated kernels $\bar{k}_n$ and $\bar{h}_n{}^m$ are defined by

$$\begin{aligned} \bar{k}_1(E) &\equiv \nu r_f(E), \\ \bar{k}_n(E) &\equiv \frac{r_{et}(E)}{E(1-\alpha)} \int_{\alpha E}^{E} \bar{k}_{n-1}(E')dE', \qquad n \geqslant 2, \end{aligned} \tag{4.23}$$

and

$$\begin{aligned} \bar{h}_1{}^m(E) &\equiv r_{in}^m(E), \\ \bar{h}_n{}^m(E) &\equiv \frac{r_{et}(E)}{E(1-\alpha)} \int_{\alpha E}^{E} \bar{h}_{n-1}^m(E')dE', \qquad n \geqslant 2. \end{aligned} \tag{4.24}$$

The constants which appear in Equation 4.22 are given by

$$f_n{}^m \equiv \int_0^\infty dEP^m(E)\bar{k}_n(E), \qquad g_n{}^{m'm} \equiv \int_0^\infty dEP^{m'}(E)\bar{h}_n{}^m(E). \tag{4.25}$$

Equation 4.22 is a formal solution of Equation 4.19; no approximations were made. In order to obtain a more tractable solution, it is necessary to approximate the iterated kernels of Equations 4.23 and 4.24 in such a manner that the infinite series of Equation 4.22 may be summed. Approximations to these iterated kernels, which involve defining effective values of the elastic scattering, inelastic scattering, and fission probabilities, may be motivated by physical arguments and given mathematical form by employing continuous slowing down theory. These approximations are developed in Appendix E. Equation 4.22 is constructed from physical arguments in this appendix, also. The approximated iterated kernels have the form

$$\begin{aligned} \bar{k}_n(E) &\Rightarrow \widetilde{\nu r}_f(E) \cdot (\tilde{r}_{et}(E))^{n-1}, \qquad \text{all } n \geqslant 1, \\ \bar{h}_n{}^m(E) &\Rightarrow \tilde{r}_{in}^m(E) \cdot (\tilde{r}_{et}(E))^{n-1}, \qquad \text{all } n \geqslant 1, \end{aligned} \tag{4.26}$$

where the tilde denotes an effective probability.

When the approximated kernels are used, the sums in Equation 4.22

can be performed analytically:

$$k(E) \equiv \sum_{n=1}^{\infty} \bar{k}_n(E) \Rightarrow \widetilde{\nu r}_f(E) \cdot \sum_{n=0}^{\infty} (\tilde{r}_{et}(E))^n = \frac{\widetilde{\nu r}_f(E)}{1 - \tilde{r}_{et}(E)}, \quad (4.27)$$

$$h^m(E) \equiv \sum_{n=1}^{\infty} \bar{h}_n^m(E) \Rightarrow \tilde{r}_{in}^m(E) \cdot \sum_{n=0}^{\infty} (\tilde{r}_{et}(E))^n = \frac{\tilde{r}_{in}^m(E)}{1 - \tilde{r}_{et}(E)}, \quad (4.28)$$

$$f^m \equiv \sum_{n=1}^{\infty} f_n^m = \int_0^{\infty} dEP^m(E) \sum_{n=1}^{\infty} \bar{k}_n(E) \Rightarrow \int_0^{\infty} dEP^m(E)\left[\frac{\widetilde{\nu r}_f(E)}{1 - \tilde{r}_{et}(E)}\right], \quad (4.29)$$

$$g^{m'm} \equiv \sum_{n=1}^{\infty} g_n^{m'm} = \int_0^{\infty} dEP^{m'}(E) \sum_{n=1}^{\infty} \bar{h}_n^m(E) \Rightarrow \int_0^{\infty} dEP^{m'}(E)\left[\frac{\tilde{r}_{in}^m(E)}{1 - \tilde{r}_{et}(E)}\right]. \quad (4.30)$$

With these definitions, the importance function becomes

$$\phi^*(E) = \left[\frac{\tilde{\nu}r_f(E)}{1 - \tilde{r}_{et}(E)}\right] + \sum_{m=1}^{M} f^m\left[\frac{\tilde{r}_{in}^m(E)}{1 - \tilde{r}_{et}(E)} + \sum_{m'=1}^{M} \frac{\tilde{r}_{in}^{m'}(E) g^{m'm}}{1 - \tilde{r}_{et}(E)} + \sum_{m''=1}^{M} \sum_{m'=1}^{M} \frac{\tilde{r}_{in}^{m'}(E)}{1 - \tilde{r}_{et}(E)} g^{m'm''} g^{m''m} + \cdots\right]. \quad (4.31)$$

Once the effective probabilities are obtained, it is only necessary to evaluate the integrals of Equations 4.29 and 4.30 and perform a few multiplications in order to obtain the importance function.

## 4.3 Application to Highly Enriched Assemblies

As an application of the previous prescriptions, the spectrum and importance function have been calculated for two highly enriched assemblies. GODIVA (93.8% $U^{235}$) and JEZEBEL ($Pu^{239}$) are bare, spherical assemblies composed almost entirely of fissile nuclei.[26] In these assemblies, the moderation caused by elastic collisions may be neglected, and the inelastic scattering from all nuclei present can be described by a single value of the parameter $T$. The expressions for the spectrum and importance function take a particularly simple form in this case.

$$F(E) = \left[\frac{2}{\sqrt{\pi}} \frac{E^{1/2}}{\theta^{3/2}} e^{-E/\theta}\right] + \beta\left[\frac{Ee^{-E/T}}{T^2}\right], \quad (4.32)$$

where

$$\beta = \frac{\int_0^\infty dE r_{in}(E)\chi(E)}{1 - \int_0^\infty dE r_{in}(E)P(E)}, \tag{4.33}$$

and

$$\phi^*(E) = \nu r_f(E) + \gamma r_{in}(E), \tag{4.34}$$

where

$$\gamma = \frac{\int_0^\infty dE \nu r_f(E)P(E)}{1 - \int_0^\infty dE r_{in}(E)P(E)}, \tag{4.35}$$

and $\chi$ and $P$ are the fission and inelastic spectra, respectively.

A leakage cross section is defined from asymptotic transport theory.[51]

$$\Sigma_L = \Sigma_{tr}\left[\frac{(B/\Sigma_{tr})}{\tan^{-1}(B/\Sigma_{tr})} - 1\right], \quad \frac{B}{\Sigma_{tr}} = \frac{\pi}{R\Sigma_{tr} + d_e}, \tag{4.36}$$

where $d_e$ is the extrapolation length in units of transport mean free paths, and $R$ is the radius.

Calculations were performed for three different values of the parameter $T$ which characterizes the inelastic spectrum. There is some ambiguity in the correct definition of $T$, and the values chosen may be motivated by a variety of theoretical and experimental arguments. The three values, $T = 0.29$, 0.33, and 0.39 Mev, seem to span the range of possible values, however. This is discussed in Appendix D.

Spectral integrals, the importance function, spectral indices, and $k_{eff}$ calculated with the Equations 4.32 through 4.36 are compared with experimental data and $S_4$ 6-group calculations in Tables 4.1 through 4.6. The criticality constant $k_{eff}$ was calculated from Equations 4.2 and 4.20. In evaluating Equations 4.2, 4.20, 4.33, and 4.35, the integrals were replaced by finite sums over intervals within which the nuclear properties were assumed to have constant effective values. These effective values of the nuclear properties were evaluated from the same multigroup cross-section data[28] that were employed in the $S_4$ calculation.

Spectra calculated with the foregoing expressions are somewhat harder than central spectra predicted by the $S_4$ 6-group calculation (Tables 4.1 and 4.2). This may be due to the neglect of the elastic moderation in Equations 4.32 through 4.35. The analytical spectra agree relatively well with the measured leakage spectrum from GODIVA (Table 4.1 and Figure 4.1).

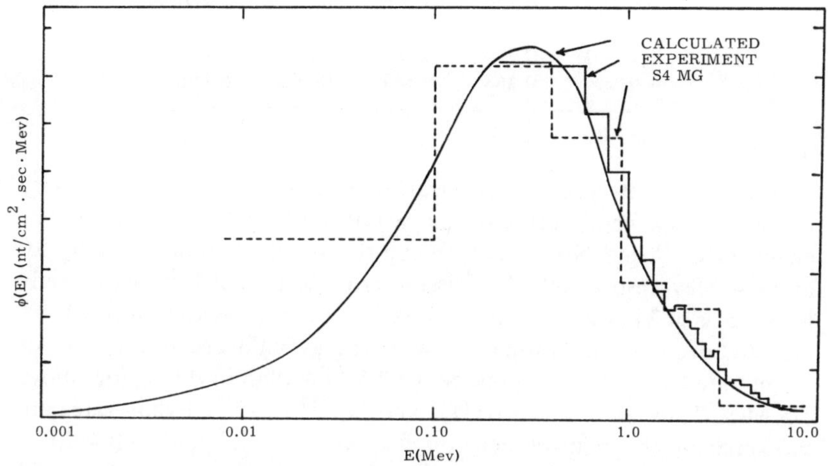


Figure 4.1 Calculated and experimental[61] spectra—GODIVA.

TABLE 4.1 *Central Flux Integrals* $\int_{\Delta E} \phi dE$ *—GODIVA*

| $\Delta E$(Mev) | $T$ (Mev) | | | exp† | $S_4$† |
|---|---|---|---|---|---|
| | 0.29 | 0.33 | 0.39 | | |
| 0–0.1 | 0.027 | 0.023 | 0.020 | 0.020 | 0.030 |
| 0.1–0.4 | 0.193 | 0.177 | 0.158 | 0.160 | 0.198 |
| 0.4–0.9 | 0.263 | 0.264 | 0.260 | 0.274 | 0.255 |
| 0.9–1.4 | 0.151 | 0.160 | 0.172 | 0.157 | 0.151 |
| 1.4–2.0 | 0.230 | 0.241 | 0.254 | 0.246 | 0.239 |
| 3.0–∞ | 0.136 | 0.135 | 0.135 | 0.143 | 0.127 |

TABLE 4.2 *Central Flux Integrals* $\int_{\Delta E} \phi dE$—*JEZEBEL*

| | $T$ (Mev) | | | |
|---|---|---|---|---|
| $\Delta E$(Mev) | 0.29 | 0.33 | 0.39 | $S_4$† |
| 0–0.1 | 0.022 | 0.020 | 0.018 | 0.026 |
| 0.1–0.4 | 0.152 | 0.143 | 0.132 | 0.159 |
| 0.4–0.9 | 0.226 | 0.226 | 0.224 | 0.226 |
| 0.9–1.4 | 0.157 | 0.162 | 0.169 | 0.157 |
| 1.4–3.0 | 0.275 | 0.281 | 0.290 | 0.276 |
| 3.0–∞ | 0.168 | 0.167 | 0.168 | 0.156 |

† See Reference 26.

In order to interpret the spectral index comparison, it is necessary to review the general nature of the various cross sections involved. The elements $U^{238}$ and $Np^{237}$ have fission thresholds at approximately 1.0 and 0.4 Mev, respectively. The fission cross section of $U^{235}$ varies only by a factor of 2 in the range of interest. The $(n, \gamma)$ cross sections for $U^{238}$ and Au decrease monotonically with energy, and are two orders of magnitude larger in the region below 0.5 Mev than in the region above 1.5 Mev. Thus, the ratios of $U^{238}$ and $Np^{237}$ to $U^{235}$ fission rates are measures of the relative neutron flux above ~1.0 and ~0.4 Mev, respectively. The ratio of the capture rate to the fission rate for $U^{238}$ is a gross measure of the ratio of the low energy flux to the high energy flux. The ratio of the Au capture rate to the $U^{235}$ fission rate is a measure of the low energy flux, while the ratio of $Np^{237}$ and $U^{238}$ fission rates relates the relative flux above ~0.4 Mev to the relative flux above ~1.0 Mev.

Spectral indices evaluated with the analytical spectra agree with measured values to within 10%, except for the Au/$U^{235}$ ratio in JEZEBEL (Tables 4.3 and 4.4). This agreement is at least as good as that obtained with the $S_4$ calculation, in general. The spectral indices are relatively insensitive to the value of T. Consequently, certain gross attributes of the spectrum, as indicated in the preceding paragraph, are predicted in accord with experiment.

The importance function of Equation 4.34 is compared with the central importance function of the $S_4$ calculation for GODIVA in Table 4.5. This function is relatively insensitive to $T$, and in substantial agreement with the $S_4$ calculation.

Values of $k_{eff}$ were calculated as the reciprocal ratio of $\nu_c$, which

TABLE 4.3 *Central Spectral Indices—GODIVA*

| | $T$ (Mev) | | | | |
|---|---|---|---|---|---|
| | 0.29 | 0.33 | 0.39 | exp‡ | $S_4$† |
| $\bar{\sigma}_f^{25}/\bar{\sigma}_f^{28}$ | 6.35 | 6.18 | 5.91 | 5.94–6.59 | 6.42 |
| $\bar{\sigma}_f^{37}/\bar{\sigma}_f^{28}$ | 5.57 | 5.60 | 5.57 | 4.8 –5.2 | 5.58 |
| $\bar{\sigma}_f^{37}/\bar{\sigma}_f^{25}$ | 0.877 | 0.907 | 0.943 | 0.78–0.88 | 0.869 |
| $\bar{\sigma}_{n,\gamma}^{28}/\bar{\sigma}_f^{28}$ | 0.53 | 0.51 | 0.48 | 0.47 | 0.54 |
| $\bar{\sigma}_{n,\gamma}^{Au}/\bar{\sigma}_f^{25}$ | 0.103 | 0.101 | 0.097 | 0.094 | 0.105 |

TABLE 4.4 *Central Spectral Indices—JEZEBEL*

| | $T$ (Mev) | | | | |
|---|---|---|---|---|---|
| | 0.29 | 0.33 | 0.39 | exp‡ | $S_4$† |
| $\bar{\sigma}_f^{25}/\bar{\sigma}_f^{28}$ | 5.21 | 5.13 | 5.02 | 4.9 | 5.39 |
| $\bar{\sigma}_f^{37}/\bar{\sigma}_f^{28}$ | 5.03 | 5.05 | 5.05 | 4.6 | 5.11 |
| $\bar{\sigma}_f^{37}/\bar{\sigma}_f^{25}$ | 0.966 | 0.984 | 1.005 | 0.99 | 0.949 |
| $\bar{\sigma}_{n,\gamma}^{28}/\bar{\sigma}_f^{28}$ | 0.40 | 0.39 | 0.38 | | 0.423 |
| $\bar{\sigma}_{n,\gamma}^{Au}/\bar{\sigma}_f^{25}$ | 0.093 | 0.092 | 0.089 | 0.077 | 0.096 |

† Based on spectral integrals reported in Reference 26 and detector cross sections reported in Reference 28.

‡ See References 26 and 27.

TABLE 4.5 *Central Importance Function—GODIVA*

| | $T$ (Mev) | | | |
|---|---|---|---|---|
| $\Delta E$(Mev) | 0.29 | 0.33 | 0.39 | $S_4$† |
| 0–0.1 | 1.334 | 1.334 | 1.334 | 1.431 |
| 0.1–0.4 | 1.054 | 1.051 | 1.049 | 1.131 |
| 0.4–0.9 | 0.952 | 0.949 | 0.947 | 0.949 |
| 0.9–1.4 | 0.968 | 0.964 | 0.961 | 0.951 |
| 1.4–3.0 | 1.004 | 1.000 | 0.997 | 0.975 |
| 3.0–∞ | 1.084 | 1.080 | 1.077 | 1.039 |

† See Reference 20.

TABLE 4.6 $k_{eff}$

| | $T$ (Mev) | | |
|---|---|---|---|
| | 0.29 | 0.33 | 0.39 |
| GODIVA (Equation 4.2) | 1.014 | 1.011 | 1.007 |
| GODIVA (Equation 4.20) | 1.014 | 1.011 | 1.008 |
| JEZEBEL (Equation 4.2) | 1.159 | 1.160 | 1.160 |
| JEZEBEL (Equation 4.20) | 1.159 | 1.160 | 1.160 |

satisfies Equation 4.2 or 4.20, to the actual value of $\nu$ (Table 4.6). The criticality of GODIVA, which is on the order of 6 transport mean free paths in diameter, was predicted to within 1.5%. The poorer results obtained for JEZEBEL, which is only on the order of 2 mean free paths in diameter, are probably attributable to the invalid use of asymptotic transport theory to represent the leakage. The constant $k_{eff}$ is relatively insensitive to $T$, varying by less than 1% for variations of $T$ between 0.29 and 0.39 Mev.

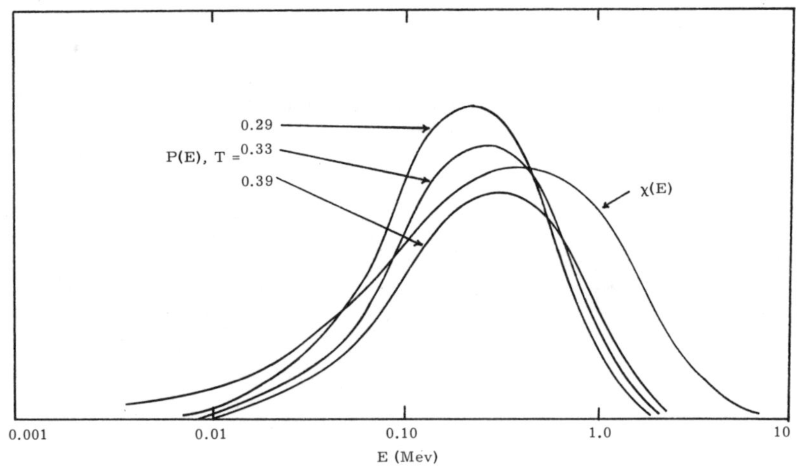


Figure 4.2 Inelastic and fission spectra.

In order to illustrate the components of the spectrum and the effect of $T$ upon the inelastic spectra, the fission spectrum and the three inelastic spectra are plotted in Figure 4.2.

## 4.4 Application to Dilute Fast Assemblies

A more practical application of the expressions that have been developed for the spectrum and importance function is to assemblies that are diluted with structural and coolant materials. The analytical expressions have been used to evaluate certain integral properties of four dilute Zero-Power-Reactor (ZPR) assemblies,[22,29,38,57] which are typical of cores being considered for commercial development. These assemblies were diluted with stainless steel and either sodium or aluminum. In two cases, carbon was present in the form of carbide fuel elements. Compositions are specified for the four assemblies in Table 4.7.

TABLE 4.7 *Dilute ZPR Assemblies Composition (v/o)*

| Assembly | $U^{235}$ | $U^{238}$ | C | Al | Na | SS |
|---|---|---|---|---|---|---|
| ZPR-III, 6F | 0.1400 | 0.1590 | | 0.3140 | | 0.1241 |
| ZPR-III, 31 | 0.0581 | 0.0914 | | 0.2350 | | 0.2450 |
| ZPR-III, 34 | 0.0467 | 0.1030 | 0.0908 | 0.2550 | | 0.2483 |
| ZPR-VI, 2 | 0.0476 | 0.2052 | 0.1540 | | 0.3595 | 0.1751 |

In treating inelastic scattering, the inelastic spectra of $U^{235}$ and $U^{238}$ were assumed to be characterized by the same parameter $T_1$, and the inelastic spectra of all dilutants were assumed to be characterized by a different parameter $T_2$. Calculations were performed for two values of $T_1$ (0.16 and 0.26 Mev), chosen to span the range of theoretically and experimentally motivated values, and for $T_2 = 1.5$ Mev, which was estimated from measured excitation functions for the principal dilutants (see Appendix D).

Equations 4.15 and 4.31, with $M = 2$, were evaluated for the material properties listed in Table 4.7. Because the assemblies were many mean free paths in dimension, leakage was neglected. Integrals involved in the definition of $\beta^m$ were replaced by finite sums and evaluated with data from the ANL-635 multigroup cross-section set. This set is a modified version[12,13] of the earlier YOM set.[68] To simplify the calculation, elastic scattering by the fuel was neglected.

In these assemblies, $(1 - \alpha)$ is on the order of 0.1, so that the neglected terms in the expression for the collision density are on the order of 0.01. The series for $\beta^m$, Equation 4.9, converged in about 15 terms, and the series for $\phi^*$, Equation 4.31, converged in about 8 terms. A convergence criterion of 0.01 was imposed. It should be noted that evaluation of the second and higher terms in these series involves nothing more than the multiplication of previously evaluated quantities.

TABLE 4.8 *ZPR Central Flux Integrals*—$\int_{\Delta E} \phi dE$

| | ZPR-III, 6F | | | ZPR-III, 31 | | | ZPR-III, 34 | | ZPR-VI, 2 |
|---|---|---|---|---|---|---|---|---|---|
| $\Delta E$(Mev) | $T_1 = .16$ | $T_1 = .26$ | $S_4$† | $T_1 = .16$ | $T_1 = .26$ | $S_4$† | $T_1 = .16$ | $S_4$† | $T_1 = .16$ |
| ∞–3.668 | 0.034 | 0.032 | 0.036 | 0.024 | 0.027 | 0.022 | 0.026 | 0.018 | 0.025 |
| 3.668–2.225 | 0.050 | 0.047 | 0.067 | 0.039 | 0.037 | 0.044 | 0.037 | 0.038 | 0.033 |
| 2.225–1.35 | 0.065 | 0.067 | 0.102 | 0.052 | 0.052 | 0.077 | 0.049 | 0.067 | 0.042 |
| 1.35–0.825 | 0.086 | 0.124 | 0.143 | 0.075 | 0.102 | 0.123 | 0.070 | 0.107 | 0.056 |
| 0.825–0.5 | 0.118 | 0.191 | 0.183 | 0.103 | 0.175 | 0.174 | 0.082 | 0.148 | 0.066 |
| 0.5–0.3 | 0.178 | 0.212 | 0.173 | 0.167 | 0.220 | 0.179 | 0.144 | 0.155 | 0.145 |
| 0.3–0.18 | 0.191 | 0.158 | 0.125 | 0.208 | 0.184 | 0.139 | 0.209 | 0.132 | 0.222 |
| 0.18–0.11 | 0.135 | 0.090 | 0.081 | 0.161 | 0.107 | 0.096 | 0.173 | 0.102 | 0.185 |
| 0.11–0.067 | 0.078 | 0.045 | 0.052 | 0.095 | 0.056 | 0.068 | 0.105 | 0.082 | 0.118 |
| 0.067–0.0407 | 0.037 | 0.019 | 0.028 | 0.048 | 0.026 | 0.049 | 0.062 | 0.069 | 0.060 |
| 0.0407–0.025 | 0.021 | 0.011 | 0.006 | 0.018 | 0.009 | 0.015 | 0.025 | 0.033 | 0.027 |
| 0.025–0.015 | 0.006 | 0.003 | 0.003 | 0.007 | 0.004 | 0.011 | 0.014 | 0.028 | 0.016 |
| 0.015–0.0091 | 0.001 | 0.001 | 0.001 | 0.002 | 0.001 | 0.002 | 0.003 | 0.011 | 0.004 |
| 0.0091–0.0055 | — | — | — | 0.001 | — | 0.001 | 0.001 | 0.005 | 0.001 |
| 0.0055–0.0021 | — | — | — | — | — | — | — | 0.004 | — |
| 0.0021–0.0005 | — | — | — | — | — | — | — | 0.001 | — |

† See Reference 12.

Flux integrals, spectral indices, the importance function, and reactivity worths calculated with the analytical formulas are compared with measurements and $S_4$ calculations at the center of the assemblies.[12,13,38,40,57] The $S_4$ calculations employed 16 energy groups and used the same nuclear data that were used to evaluate the analytical formulas.

Central flux integrals calculated with Equation 4.15 are compared with 16-group calculations in Table 4.8. Two features should be noted:

1. The analytical spectrum has fewer neutrons above ~1 Mev than the multigroup spectrum. The distribution of this "fast" flux is flatter for the analytical spectrum than for the multigroup spectrum.
2. The analytical spectrum is more sharply peaked than the multigroup spectrum. The location of the peak in the analytical spectrum is sensitive to the parameter $T_1$, and is generally at lower energies than the peak in the multigroup spectrum.

A more pertinent comparison is between reaction rates calculated from the analytical spectra and measured reaction rates. Spectral indices, which are ratios of reaction rates, have been measured at the center of the four ZPR assemblies considered in this section.

The elements $U^{236}$ and $U^{238}$ have threshold fission cross sections which vanish below ~1 Mev, while the threshold of the $U^{234}$ and $Pu^{240}$ fission cross sections is on the order of 0.3 Mev; $U^{235}$, $U^{233}$, and $Pu^{239}$ have nonthreshold fission cross sections which vary within an order of magnitude over the range of interest. The $Pu^{239}$ fission cross section is relatively flat in the range $10^4 - 10^7$ ev, while $U^{233}$ and $U^{235}$ have relatively flat fission cross sections in the range $10^5 - 10^7$ ev. The $Pu^{239}$ fission cross section is less than the $U^{235}$ fission cross section below $10^5$ ev, at which point it becomes larger. The $U^{233}$ fission cross section is roughly 1 barn larger than the $U^{235}$ fission cross section and has approximately the same variation with energy. The $(n, \gamma)$ cross section of $U^{238}$ decreases monotonically with energy and varies by two orders of magnitude between the kev and Mev regions.

Thus, the ratio of $U^{236}$ or $U^{238}$ to $U^{235}$ fission rates is a measure of the relative flux above 1 Mev, while the $U^{234}$ or $Pu^{240}$ to $U^{235}$ fission rate ratio indicates the relative flux above 0.3 Mev. These ratios naturally increase as the neutron flux is shifted to higher energies. The ratio of the $U^{238}$ capture rate or the $Pu^{239}$ fission rate to the $U^{235}$ fission rate is sensitive to the low energy part of the spectrum. The latter decreases and the former increases as the neutron flux is shifted to higher energies. The $U^{233}$ to $U^{235}$ fission rate ratio is relatively insensitive to the spectrum.

TABLE 4.9 *Central Spectral Indices in Dilute ZPR Assemblies*

| | T (Mev) | | | | |
|---|---|---|---|---|---|
| | 0.16 | 0.26 | exp† | $S_4$† | |
| $\bar{\sigma}_f^{26}/\bar{\sigma}_f^{25}$ | 0.115 | 0.132 | | 0.174 | |
| $\bar{\sigma}_f^{28}/\bar{\sigma}_f^{25}$ | 0.055 | 0.057 | 0.069 | 0.078 | |
| $\bar{\sigma}_f^{40}/\bar{\sigma}_f^{25}$ | 0.334 | 0.431 | | 0.508 | |
| $\bar{\sigma}_f^{24}/\bar{\sigma}_f^{25}$ | 0.333 | 0.431 | | 0.498 | ZPR-III, 6F |
| $\bar{\sigma}_f^{23}/\bar{\sigma}_f^{25}$ | 1.567 | 1.587 | | 1.570 | |
| $\bar{\sigma}_f^{49}/\bar{\sigma}_f^{25}$ | 1.215 | 1.274 | | 1.288 | |
| $\bar{\sigma}_c^{28}/\bar{\sigma}_f^{25}$ | 0.114 | 0.107 | 0.084 | 0.103 | |
| $\bar{\sigma}_f^{26}/\bar{\sigma}_f^{25}$ | 0.089 | 0.105 | 0.112 | 0.120 | |
| $\bar{\sigma}_f^{28}/\bar{\sigma}_f^{25}$ | 0.041 | 0.045 | 0.047 | 0.052 | |
| $\bar{\sigma}_f^{40}/\bar{\sigma}_f^{25}$ | 0.271 | 0.361 | 0.326 | 0.394 | |
| $\bar{\sigma}_f^{24}/\bar{\sigma}_f^{25}$ | 0.272 | 0.367 | 0.347 | 0.383 | ZPR-III, 31 |
| $\bar{\sigma}_f^{23}/\bar{\sigma}_f^{25}$ | 1.565 | 1.590 | 1.523 | 1.569 | |
| $\bar{\sigma}_f^{49}/\bar{\sigma}_f^{25}$ | 1.184 | 1.249 | 1.181 | 1.230 | |
| $\bar{\sigma}_c^{28}/\bar{\sigma}_f^{25}$ | 0.119 | 0.111 | | 0.114 | |
| $\bar{\sigma}_f^{26}/\bar{\sigma}_f^{25}$ | 0.082 | | 0.080 | 0.097 | |
| $\bar{\sigma}_f^{28}/\bar{\sigma}_f^{25}$ | 0.039 | | 0.037 | 0.041 | |
| $\bar{\sigma}_f^{40}/\bar{\sigma}_f^{25}$ | 0.242 | | 0.282 | 0.311 | |
| $\bar{\sigma}_f^{24}/\bar{\sigma}_f^{25}$ | 0.241 | | 0.257 | 0.309 | ZPR-III, 34 |
| $\bar{\sigma}_f^{23}/\bar{\sigma}_f^{25}$ | 1.553 | | 1.455 | 1.545 | |
| $\bar{\sigma}_f^{49}/\bar{\sigma}_f^{25}$ | 1.156 | | 1.067 | 1.153 | |
| $\bar{\sigma}_c^{28}/\bar{\sigma}_f^{25}$ | 0.123 | | 0.122 | 0.125 | |
| $\bar{\sigma}_f^{26}/\bar{\sigma}_f^{25}$ | 0.072 | 0.079 | 0.074 | 0.086 | |
| $\bar{\sigma}_f^{28}/\bar{\sigma}_f^{25}$ | 0.034 | 0.037 | 0.036 | 0.038 | |
| $\bar{\sigma}_f^{40}/\bar{\sigma}_f^{25}$ | 0.207 | 0.283 | 0.254 | 0.273 | |
| $\bar{\sigma}_f^{24}/\bar{\sigma}_f^{25}$ | 0.208 | 0.292 | 0.225–0.233 | 0.267 | ZPR-VI, 2 |
| $\bar{\sigma}_f^{23}/\bar{\sigma}_f^{25}$ | 1.552 | 1.586 | 1.45–1.52 | 1.532 | |
| $\bar{\sigma}_f^{49}/\bar{\sigma}_f^{25}$ | 1.140 | 1.210 | 1.04–1.09 | 1.113 | |
| $\bar{\sigma}_c^{28}/\bar{\sigma}_f^{25}$ | 0.125 | 0.116 | 0.106 | 0.128 | |

† See References 12, 40, 57.

Based on this interpretation of the spectral indices, several points emerge from Table 4.9:

1. The analytical spectrum above ~1 Mev is relatively insensitive to $T_1$, and agrees with experiment relatively well, except for the moderately dilute assembly 6F, wherein the analytical spectrum underpredicts the "fast" flux ($\bar{\sigma}_f^{26}/\bar{\sigma}_f^{25}$ and $\bar{\sigma}_f^{28}/\bar{\sigma}_f^{25}$).

TABLE 4.10 *ZPR Central Importance Functions*

| | ZPR-III, 31 | | | ZPR-III, 34 | |
|---|---|---|---|---|---|
| | $T = 0.16$ Mev | $T = 0.26$ Mev | $S_4$† | $T = 0.26$ Mev | $S_4$† |
| ∞–3.668 Mev | 0.073 | 0.074 | 0.061 | 0.071 | 0.063 |
| 3.668–2.225 Mev | 0.073 | 0.075 | 0.061 | 0.072 | 0.062 |
| 2.225–1.35 Mev | 0.072 | 0.073 | 0.058 | 0.068 | 0.060 |
| 1.35–0.825 Mev | 0.067 | 0.069 | 0.053 | 0.062 | 0.055 |
| 0.825–0.5 Mev | 0.065 | 0.068 | 0.053 | 0.062 | 0.056 |
| 0.5–0.3 Mev | 0.066 | 0.068 | 0.056 | 0.062 | 0.058 |
| 0.3–0.18 Mev | 0.064 | 0.065 | 0.058 | 0.062 | 0.060 |
| 0.18–0.11 Mev | 0.063 | 0.063 | 0.061 | 0.062 | 0.061 |
| 0.11–0.067 Mev | 0.062 | 0.061 | 0.062 | 0.064 | 0.062 |
| 0.067–0.0407 Mev | 0.060 | 0.059 | 0.064 | 0.064 | 0.063 |
| 0.0407–0.025 Mev | 0.060 | 0.058 | 0.065 | 0.062 | 0.065 |
| 0.25–0.015 Mev | 0.055 | 0.053 | 0.066 | 0.059 | 0.066 |
| 0.015–0.0091 Mev | 0.057 | 0.055 | 0.071 | 0.059 | 0.069 |
| 0.0091–0.0055 Mev | 0.056 | 0.054 | 0.071 | 0.059 | 0.068 |
| 0.0055–0.0021 Mev | 0.055 | 0.054 | 0.072 | 0.059 | 0.068 |
| 0.0021–0.0005 Mev | 0.052 | 0.051 | 0.068 | 0.053 | 0.064 |

† See Reference 12.

2. The analytical spectrum in the region $0.3 < E < 1.0$ Mev is quite sensitive to the choice of $T_1$. Values of $T_1 = 0.16$ and 0.26 Mev underpredict and overpredict, respectively, the relative flux in this region ($\bar{\sigma}_f^{24}/\bar{\sigma}_f^{25}$ and $\bar{\sigma}_f^{40}/\bar{\sigma}_f^{25}$).
3. The relative flux in the low energy region is slightly overpredicted by the analytical spectrum ($\bar{\sigma}_c^{28}/\bar{\sigma}_f^{25}$), although the relative flux above $10^5$ ev is overpredicted by the analytical spectrum ($\bar{\sigma}_f^{49}/\bar{\sigma}_f^{25}$).
4. Spectral indices calculated with the analytical spectrum agree with experiment as well as those resulting from the multigroup calculation, in general.

Importance functions evaluated from Equation 4.31 are compared with the central importance function of the $S_4$ 16-group calculation in Table 4.10. The analytical importance function is larger at high energies and smaller at low energies than the multigroup importance function. The multigroup importance function has a pronounced dip around 1 Mev which is not present in the analytical importance function. The analytical importance function is relatively insensitive to $T_1$.

Measured central reactivity worths provide an indirect means of comparing the importance function obtained from Equation 4.31 with experiment. The perturbation theory expression for the reactivity worth of a small sample is

$$\delta k = \frac{\int_0^\infty dE\phi^*(E)\delta M \cdot \phi(E)}{\int_0^\infty dE\phi^*(E)\phi(E)}, \tag{4.37}$$

TABLE 4.11 *Central Reactivity Worths in Dilute ZPR-III Assemblies*

| | Assembly 34 | | | |
|---|---|---|---|---|
| | $T_1 = 0.16$ | | expt† | $S_4$† |
| $\delta k^{25}$(barn/atom) | 2.093 | | $1.961 \pm 0.035$ | 1.896 |
| $\delta k^{28}$(barn/atom) | −0.111 | | $-0.108 \pm 0.003$ | −0.094 |
| $\delta k^{25}/\delta k^{28}$ | −18.83 | | −18.16 | −20.17 |
| | **Assembly 31** | | | |
| | $T_1 = 0.16$ | $T_1 = 0.26$ | expt† | $S_4$† |
| $\delta k^{25}$(barn/atom) | 2.139 | 2.045 | $1.837 \pm 0.035$ | 1.820 |
| $\delta k^{28}$(barn/atom) | −0.091 | −0.064 | $-0.050 \pm 0.003$ | −0.040 |
| $\delta k^{25}/\delta k^{28}$ | −23.51 | −31.95 | −36.74 | −45.50 |

† See Reference 12.

where $\delta M$ is the change in the net neutron production operator (production minus destruction) caused by the introduction of a small sample. Equation 4.37 was evaluated with the analytical spectra and importance functions, and is compared with measured values, and values calculated from Equation 4.37 with multigroup spectra and importance functions, in Table 4.11. The analytical spectra and importance functions lead to reactivity worths that agree with experiment as well as the multigroup reactivity worths, in general.

## 4.5 Conclusions

Certain general conclusions may be drawn from the results of the preceding two sections.

1. The analytical formulas for the spectrum and importance function may be used to calculate integral parameters such as reaction rates and reactivity worths for a wide range of fast-reactor compositions. Integral parameters so calculated agree with measured values as well as parameters calculated with a multigroup model, in general.
2. Theoretical and experimental considerations may be invoked to specify a range of reasonable values of the parameter $T$, which characterizes the inelastic spectrum. In general, the integral parameters of the assembly are relatively insensitive to the choice of $T$.
3. Representation of the inelastic spectrum by the evaporation model is adequate, as demonstrated by the calculations for the highly enriched assemblies, in which elastic scattering has a negligible effect upon the spectrum.

Because preliminary calculations are usually intended to survey certain integral parameters such as critical mass, conversion ratio, and reactivity coefficients, the analytical expressions developed in this chapter can provide a simple means of making these calculations. This is particularly true if these expressions are used to construct modes and weighting functions for the energy modal calculation, because then the spectrum at any point is determined as a combination of the several modes, and there is no *a priori* demand for an exact calculation of a given mode.

# 5. Applications of Modal Diffusion Approximations to Fast-Reactor Assemblies

One of the purposes of this monograph is the development of simple calculational models that can be used to predict certain neutronic properties required in preliminary or survey studies of fast-reactor assemblies. In this chapter, it is demonstrated that the modal diffusion approximations, which were developed in Section 3.2, are suitable for this purpose.

The direct method (Section 3.2.1), whereby the neutron balance equation is reduced to a system of algebraic equations, is applied to calculate the criticality and also certain reaction rates in GODIVA, which is a highly enriched $U^{235}$ assembly. Neutronic properties of the dilute ZPR-III, Assembly 39, which is typical of cores considered for commercial development, are calculated with the semidirect energy modal (Section 3.2.3) and space modal (Section 3.2.2) approximations.

## 5.1 Direct Method

GODIVA[26] is a highly enriched (93.8% $U^{235}$), bare, spherical assembly of radius ~8.7 cm. When the leakage is treated with asymptotic transport theory by defining a leakage cross section,

$$\Sigma_L(E) = \Sigma_{tr}(E)\left[\frac{B/\Sigma_{tr}(E)}{\tan^{-1}(B/\Sigma_{tr}(E))} - 1\right], \qquad \frac{B}{\Sigma_{tr}} = \frac{\pi}{R\Sigma_{tr}(E) + d_e}, \tag{5.1}$$

and the effect of elastic moderation is neglected, the total cross section is the sum of the nonelastic and the leakage cross sections

$$\Sigma(E) = \Sigma_{ne}(E) + \Sigma_L(E). \tag{5.2}$$

The neutron balance equation is

$$H(E, \lambda)\phi(E) \equiv \lambda\chi(E) \int_0^\infty dE'\nu\Sigma_f(E')\phi(E') + \int_E^\infty dE'\Sigma_{in}(E'/E)\phi(E') - \Sigma(E)\phi(E) = 0. \tag{5.3}$$

The quantity $\lambda$ is a free parameter ($\lambda^{-1} = k_{eff}$) which must be chosen so that Equation 5.3 is satisfied.

Suitable modes may be chosen from physical considerations. Because elastic moderation is negligible, the neutrons in the assembly may be considered as either virgin fission neutrons or neutrons that have been inelastically scattered. Thus, the fission spectrum and the inelastic spectrum of $U^{235}$ are reasonable expansion modes for the flux.

$$\phi(E) = a_1 \cdot \frac{\chi(E)}{\Sigma(E)} + a_2 \cdot \frac{P^{25}(E)}{\Sigma(E)}. \tag{5.4}$$

For this calculation, Equation 4.3 is used for $\chi$ and Equation 4.4 with $T_{25} = 0.39$ Mev is used for $P^{25}$.

Use of the importance function as a weighting function was motivated in Section 3.3. Two linearly independent weighting functions are

$$\alpha_1(E) = \begin{cases} \phi^*(E), & E < 0.9 \text{ Mev}, \\ 0, & E > 0.9 \text{ Mev}, \end{cases} \qquad \alpha_2(E) = \begin{cases} 0, & E < 0.9 \text{ Mev}, \\ \phi^*(E), & E > 0.9 \text{ Mev}, \end{cases} \tag{5.5}$$

where $\phi^*$ can be evaluated from Equation 4.34.

When Equation 5.4 is substituted into Equation 5.3 and the resulting equation is multiplied, in turn, by each $\alpha_i$ and integrated over energy, two homogeneous algebraic equations in $a_1$ and $a_2$ are obtained:

$$\begin{aligned} \int_0^{0.9} dE\phi^*(E)H(E, \lambda)\left[a_1 \frac{\chi(E)}{\Sigma(E)} + a_2 \frac{P^{25}(E)}{\Sigma(E)}\right] &= 0, \\ \int_{0.9}^\infty dE\phi^*(E)H(E, \lambda)\left[a_1 \frac{\chi(E)}{\Sigma(E)} + a_2 \frac{P^{25}(E)}{\Sigma(E)}\right] &= 0. \end{aligned} \tag{5.6}$$

The first of these equations requires that the $a_i$ be determined so that the modal expansion satisfies the balance equation in an importance weighted integral sense over the interval $0 \leqslant E \leqslant 0.9$ Mev. The second equation requires that the $a_i$ be determined so that the modal expansion satisfies the balance equation in an importance weighted integral sense over the interval $0.9 \text{ Mev} \leqslant E < \infty$. The complementary intervals are chosen to ensure that Equations 5.6 will be computationally linearly independent. Most of the inelastic scattering takes place above 0.9 Mev, while most of the inelastic spectrum is below 0.9 Mev, which should lead to distinctly different values of the coefficients in Equations 5.6.

The free parameter $\lambda$ is determined from the requirement that the determinant of the coefficients vanishes in Equations 5.6, which is necessary in order that a nontrivial solution exist. Once $\lambda$ is determined, $a_2/a_1$ can be evaluated from either of Equations 5.6. The effective multiplication constant determined in this manner is

$$k_{eff} = \lambda^{-1} = 1.006.$$

Spectral indices calculated with Equation 5.4 are compared with experimental values and with values calculated with an $S_4$ 6-group model in Table 5.1. The relatively good agreement indicates that the modal calculation predicts relative reaction rates in reasonable accord with experiment, and implies a commensurate ability to predict conversion ratios and relative isotopic depletion rates.

In solving Equations 5.6, the integrals were replaced by finite sums and evaluated with the same nuclear data that were used in the $S_4$ calculations.[28]

In Section 4.3, Equation 5.3 was solved "exactly," i.e., by the method of successive approximations. Because the same treatment of the inelastic spectrum and the same nuclear data were used in Section 4.3 as are used in this section, the results obtained for $T = 0.39$ Mev in Section 4.3 may be used as a standard. A comparison of Tables 5.1 and 4.3, and of Table 4.6 with the effective multiplication constant just obtained, indicates that results obtained from the modal approximation agree very well with the "exact" results.

TABLE 5.1 *Central Spectral Indices—GODIVA*

| | Modal | exp‡ | $S_4$† |
|---|---|---|---|
| $\bar{\sigma}_f^{25}/\bar{\sigma}_f^{28}$ | 6.05 | 5.94–6.59 | 6.42 |
| $\bar{\sigma}_f^{37}/\bar{\sigma}_f^{28}$ | 5.57 | 4.8 –5.21 | 5.58 |
| $\bar{\sigma}_f^{37}/\bar{\sigma}_f^{25}$ | 0.949 | 0.78–0.88 | 0.869 |
| $\bar{\sigma}_{n,\gamma}^{28}/\bar{\sigma}_f^{28}$ | 0.49 | 0.47 | 0.54 |
| $\bar{\sigma}_{n,\gamma}^{Au}/\bar{\sigma}_f^{25}$ | 0.097 | 0.094 | 0.105 |

† Based on spectra reported in Reference 26 and cross sections reported in Reference 28.

‡ See References 26 and 27.

## 5.2 Semidirect Method: Energy Modes

The purpose of this section is the application of the energy modal formalism developed in Section 3.2.3 to the calculation of certain neutronic properties of a dilute fast assembly that is representative of cores being considered for commercial development. ZPR-III, Assembly 39[3] is a two-region spherical facility (Table 5.2). The core is 45.4 cm in radius and is surrounded by a 30-cm blanket of depleted uranium.

TABLE 5.2 *Composition of ZPR-III, Assembly* 39 ($v/o$)

| | $U^{235}$ | $U^{238}$ | Al | SS |
|---|---|---|---|---|
| Core | 0.0981 | 0.0914 | 0.2350 | 0.2450 |
| Blanket | 0.0019 | 0.8330 | | 0.0731 |

Two separate energy modal calculations were performed. The reasons for this were to illustrate different means of selecting modes and weighting functions, and to check the sensitivity of the predicted neutronic properties to the choice of modes and weighting functions.

The general formalism was developed in Section 3.2.3. When the energy dependence of the flux is expanded in two energy modes $\psi_i(E)$,

$$\phi(r, E) = \rho_1(r)\psi_1(E) + \rho_2(r)\psi_2(E), \tag{5.7}$$

Equation 3.30 becomes

$$\begin{bmatrix} (-D^{11}\nabla^2 + \Sigma^{11}) & (-D^{12}\nabla^2 + \Sigma^{12}) \\ (-D^{21}\nabla^2 + \Sigma^{21}) & (-D^{22}\nabla^2 + \Sigma^{22}) \end{bmatrix} \begin{bmatrix} \rho_1(r) \\ \rho_2(r) \end{bmatrix} = 0 \tag{5.8}$$

for a source-free problem. The modal coefficients $\Sigma^{ij}$ and $D^{ij}$ are defined by Equations 3.31. An equation of the form of Equation 5.8 applies to the core and another one applies to the blanket, with the modal coefficients defined in terms of the nuclear properties of the respective regions.

Appropriate boundary conditions require that the flux vanish at some extrapolated boundary $\tilde{a}$, which is taken equal to the exterior boundary of the blanket in this problem, and that the flux be everywhere finite. The interface boundary conditions are continuity of flux,

$$\begin{bmatrix} \rho_1(R_0) \\ \rho_2(R_0) \end{bmatrix}_{\text{core}} = \begin{bmatrix} \rho_1(R_0) \\ \rho_2(R_0) \end{bmatrix}_{\text{blkt}}, \tag{5.9}$$

and continuity of current,

$$\begin{bmatrix} D^{11} & D^{12} \\ D^{21} & D^{22} \end{bmatrix} \begin{bmatrix} \nabla\rho_1(R_0) \\ \nabla\rho_2(R_0) \end{bmatrix}\Bigg|_{\text{core}} = \begin{bmatrix} D^{11} & D^{12} \\ D^{21} & D^{22} \end{bmatrix} \begin{bmatrix} \nabla\rho_1(R_0) \\ \nabla\rho_2(R_0) \end{bmatrix}\Bigg|_{\text{blkt}}. \tag{5.10}$$

In Appendix B, Equation 5.8 is solved analytically, subject to the previous boundary conditions. The general solution is

| | CORE | BLKT |
|---|---|---|
| $\rho_1(r)$ | $A_1 \dfrac{\sin \alpha_+ r}{r} + C_1 \dfrac{\sinh \alpha_- r}{r}$ | $D_1 \dfrac{\sin \beta_+(\tilde{a} - r)}{r} + H_1 \dfrac{\sinh \beta_-(\tilde{a} - r)}{r},$ |
| $\rho_2(r)$ | $a_1 A_1 \dfrac{\sin \alpha_+ r}{r} + a_2 C_1 \dfrac{\sinh \alpha_- r}{r}$ | $a_3 D_1 \dfrac{\sin \beta_+(\tilde{a} - r)}{r} + a_4 H_1 \dfrac{\sinh \beta_-(\tilde{a} - r)}{r}.$ |

(5.11)

When these solutions are substituted into Equations 5.9 and 5.10, the constants $D_1$, $C_1$, and $H_1$ may be determined in terms of $A_1$. The condition for a nontrivial solution leads to a transcendental relation that may be satisfied by varying the core radius $R_0$, the blanket thickness, or the composition. These results are fully developed in Appendix B.

Reasonable energy modes $\psi_i(E)$ can be motivated from physical considerations. Because the core is on the order of 10 transport mean free paths in radius, the core may be considered as loosely coupled with respect to the spectrum (see Section 3.3.1); i.e., the core infinite medium spectrum is probably a good approximation to the actual spectrum over a significant volume of the core. The blanket influences the spectrum via those neutrons that diffuse into the blanket and there undergo inelastic and elastic moderation and absorption. Consequently, the spectrum at any point in the core or blanket might reasonably be described by some combination of two modes chosen as (1) the infinite medium spectrum of the core, and (2) the spectrum of neutrons elastically moderating, according to the properties of the blanket, from the $U^{238}$ inelastic spectrum.

These two energy modes can be constructed from the analytical expressions which were developed in Chapter 4. Mode 1, the core infinite medium spectrum, is given by Equation 4.15 divided by the total cross section, and is evaluated for the composition of the core. In Equation 4.17 $J(E)$ describes the collision density of neutrons elastically moderating from an inelastic spectrum. Thus, mode 2 is obtained from Equation 4.17 divided by the total cross section, evaluated for the blanket composition.

The inelastic parameters $T_1 = 0.26$ and $T_2 = 1.5$ Mev are used in constructing mode 1. These choices are based on the results of Chapter 4.

For mode 2, the inelastic parameter $T$ is determined from Equation D.3 for $A = 238$ and $\nu = 0.16\ \text{Mev}^{-1}$. The average energy at which neutrons in the core infinite medium spectrum have inelastic collisions with $U^{238}$ is used in place of the maximum value of the excitation energy. This results in $T = 0.1$ Mev for mode 2.

The weighting functions are chosen as

$$\alpha_1(E) = \begin{cases} \phi^*(E), & E > 0.825 \text{ Mev}, \\ 0, & E < 0.825 \text{ Mev}, \end{cases} \qquad \alpha_2(E) = \begin{cases} 0, & E > 0.825 \text{ Mev}, \\ \phi^*(E), & E < 0.825 \text{ Mev}, \end{cases}$$

where $\phi^*$ is the core infinite medium importance function and can be constructed from Equation 4.31. Because the core is loosely coupled, this should be a good approximation to the actual importance function over much of the core volume. When the weighting functions are chosen in this manner, Equations 5.8 have the physical significance of being a requirement that the modal expansion satisfies the neutron balance equation in an importance weighted integral sense over each of two complementary energy intervals. Most of the inelastic scattering and $U^{238}$ fission occurs for $E > 0.825$ Mev, while most of the $U^{235}$ fission occurs for $E < 0.825$ Mev. Thus, these weighting functions should lead to Equations 5.8 which are computationally linearly independent.

This choice of modes and weighting functions will be referred to as Set No. 1.

An alternate method of constructing modes and weighting functions involves the utilization of calculations that have been made on similar assemblies. Multigroup calculations of the central flux and adjoint have been published[12] for ZPR-III, Assembly 31, which was a cylindrical facility of approximately the same core composition as Assembly 39. Mode 1 is chosen as the central flux from this multigroup calculation, and mode 2 is the same as before. The weighting functions are the central adjoint from the multigroup calculation and a constant. In this case, Equations 5.8 have the physical significance of being requirements that the modal expansion satisfies the neutron balance equation in an importance weighted integral sense and in an unweighted integral sense. This choice of modes and weighting functions is denoted as Set No. 2. Sets No. 1 and No. 2 are listed in Table 5.3.

Numerical evaluation of the modal coefficients $D^{ij}$ and $\Sigma^{ij}$ is accomplished by replacing the integrals involved by finite sums. The ANL-635 16-group cross section set, which is a modified[12] version of the YOM set,[68] is used to define effective values of the cross sections.

The critical mass is determined by evaluating Equation B.30 for $R_0$, the critical radius. Set No. 1 and Set No. 2 lead to critical masses that are 9.2% and 12.5% larger, respectively, than the experimentally

TABLE 5.3 *Modes and Weighting Functions, Energy Modal Calculation—ZPR-III, Assembly 39*

| $\Delta E$ (Mev) | Set No. 1 | | | | Set No. 2 | | | |
|---|---|---|---|---|---|---|---|---|
| | $\psi_1$† | $\psi_2$† | $\alpha_1$ | $\alpha_2$ | $\psi_1$† | $\psi_2$† | $\alpha_1$ | $\alpha_2$ |
| ∞–3.668 | 0.027 | — | 0.074 | — | 0.022 | — | 0.061 | 0.0625 |
| 3.668–2.225 | 0.037 | — | 0.075 | — | 0.044 | — | 0.061 | 0.0625 |
| 2.225–1.35 | 0.052 | — | 0.073 | — | 0.077 | — | 0.058 | 0.0625 |
| 1.35–0.825 | 0.102 | 0.001 | 0.069 | — | 0.123 | 0.001 | 0.053 | 0.0625 |
| 0.825–0.5 | 0.175 | 0.018 | 0.068 | — | 0.174 | 0.018 | 0.053 | 0.0625 |
| 0.5–0.3 | 0.220 | 0.095 | — | 0.068 | 0.179 | 0.095 | 0.056 | 0.0625 |
| 0.3–0.18 | 0.184 | 0.188 | — | 0.065 | 0.139 | 0.188 | 0.058 | 0.0625 |
| 0.18–0.11 | 0.107 | 0.220 | — | 0.063 | 0.096 | 0.220 | 0.061 | 0.0625 |
| 0.11–0.067 | 0.056 | 0.198 | — | 0.061 | 0.068 | 0.198 | 0.062 | 0.0625 |
| 0.067–0.0407 | 0.026 | 0.159 | — | 0.059 | 0.049 | 0.159 | 0.064 | 0.0625 |
| 0.0407–0.025 | 0.009 | 0.077 | — | 0.058 | 0.015 | 0.077 | 0.065 | 0.0625 |
| 0.025–0.015 | 0.004 | 0.029 | — | 0.053 | 0.011 | 0.029 | 0.066 | 0.0625 |
| 0.015–0.0091 | 0.001 | 0.010 | — | 0.055 | 0.002 | 0.010 | 0.071 | 0.0625 |
| 0.0091–0.0055 | — | 0.004 | — | 0.054 | 0.001 | 0.004 | 0.071 | 0.0625 |
| 0.0055–0.0021 | — | 0.001 | — | 0.054 | — | 0.001 | 0.072 | 0.0625 |
| 0.0021–0.0005 | — | — | — | 0.051 | — | — | 0.068 | 0.0625 |

† These are flux integrals over $\Delta E$, i.e., $\int_{\Delta E} \psi(E)\, dE$.

loaded critical mass. In analyzing ZPR experiments,[12,13] it is conventional to "correct" the experimental critical mass to account for the effects of gaps, boundary irregularities, and heterogeneities that are not treated in the calculational model. "Corrected experimental masses" are 1% to 7% larger than actual critical masses in the ZPR-III assemblies.

The $S_4$ 16-group calculations[12,13] of critical masses in the ZPR-III assemblies differ from the "corrected experimental masses" by 0.5% to 6%. These calculations and the modal calculations employ the same nuclear data. Thus, the modal calculation of the critical mass is within the same range of accuracy as is obtained with $S_4$ 16-group calculations.

Modal combining coefficients $\rho_1$ and $\rho_2$ are plotted in Figures 5.1 and 5.2 as functions of radial position. For set No. 1 and Set No. 2, $\rho_1$, which corresponds to the core infinite medium spectrum, is dominant over most of the core, while $\rho_2$ becomes dominant in the blanket. This is consistent with the physical arguments used to motivate the choice of modes. The fact that $\rho_2$ is negative at the center of the core, in Figure 5.1, indicates that the actual spectrum there is "harder" than either mode 1 or mode 2. For Set No. 2, which employs a "harder" mode 1, both $\rho_1$ and $\rho_2$ are everywhere positive, indicating that the spectrum is intermediate between mode 1 and mode 2.

Depleted ($U^{238}$) and enriched (93.2% $U^{235}$) fission counter traverses have been measured along two radial directions in ZPR-III, Assembly 39.[3] The data are reproduced in Figures 5.3 through 5.6, where they are compared with radial fission counter traverses calculated with Equation 5.7; i.e.,

$$\bar{\sigma}_f(r) = \rho_1(r) \int_0^\infty dE\sigma_f(E)\psi_1(E) + \rho_2(r) \int_0^\infty dE\sigma_f(E)\psi_2(E). \quad (5.12)$$

All results are normalized to unity at $r = 0$. The maximum disagreement between the calculated and measured fission rates is on the order of 10%. This agreement is significant with respect to the calculation of power distributions.

Spectral indices, calculated for both sets of modes and weighting functions, are compared with values measured at the center of ZPR-III, Assembly 31, in Table 5.4. Also, $S_4$ 16-group calculations of these quantities are included.

The relative flux above ~1 Mev is predicted quite well by the energy modal and $S_4$ calculations, as may be inferred from the $U^{236}$ and $U^{238}$ fission ratios. The $U^{234}$ and $Pu^{240}$ fission ratios indicate that the relative flux above ~0.3 Mev is overpredicted by the energy modal and $S_4$

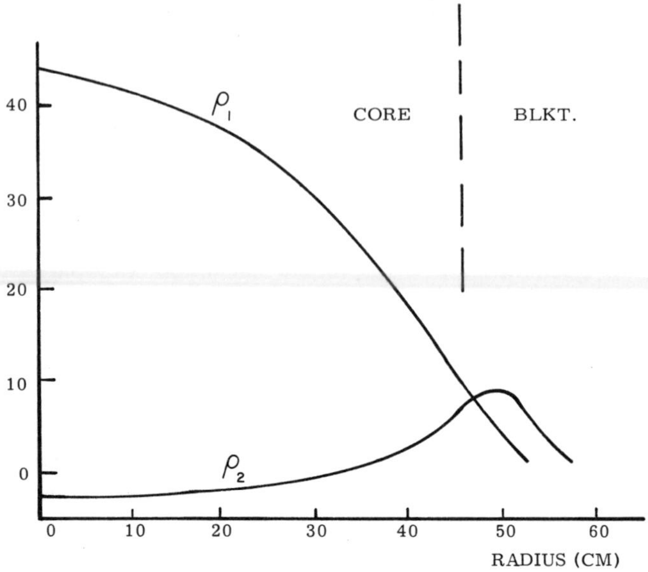


Figure 5.1 Modal combining coefficients, Set No. 1.

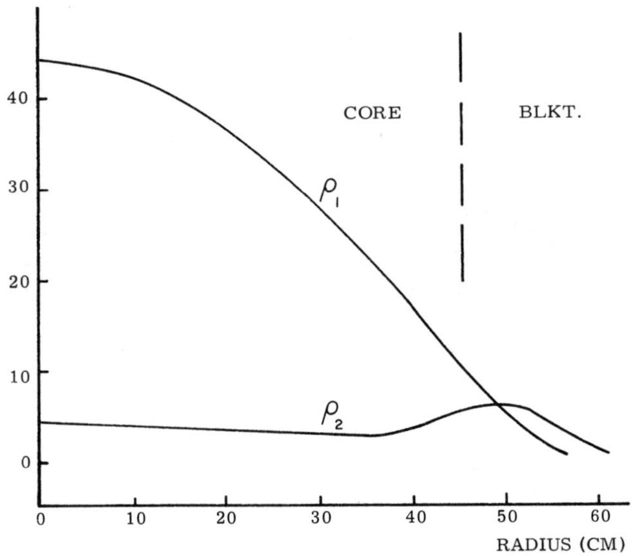


Figure 5.2 Modal combining coefficients, Set No. 2.

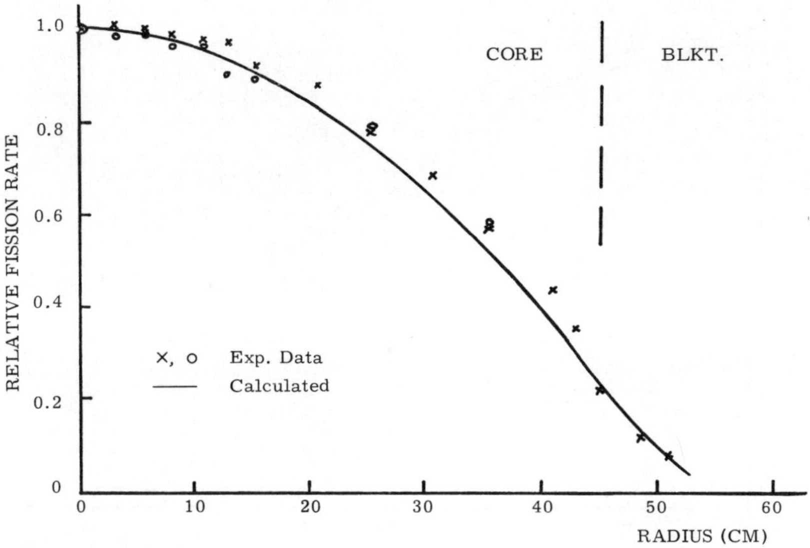


Figure 5.3 Depleted uranium fission counter radial traverse, Set No. 1, ZPR-III, Assembly 39.

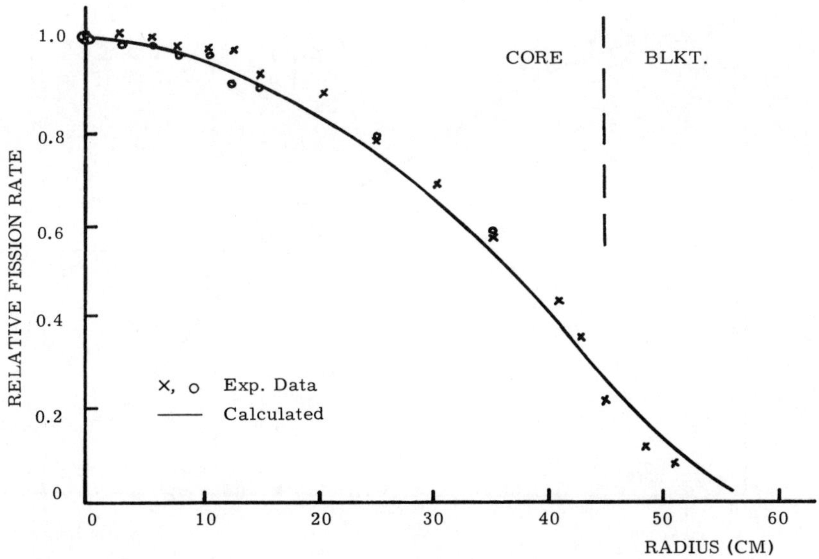


Figure 5.4 Depleted uranium fission counter radial traverse, Set No. 2, ZPR-III, Assembly 39.

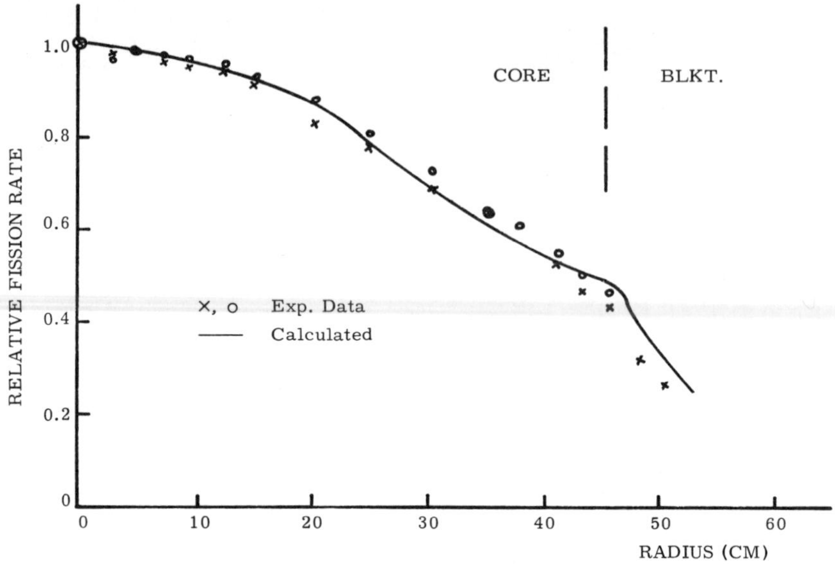


Figure 5.5 Enriched uranium fission counter radial traverse, Set No. 1, ZPR-III, Assembly 39.

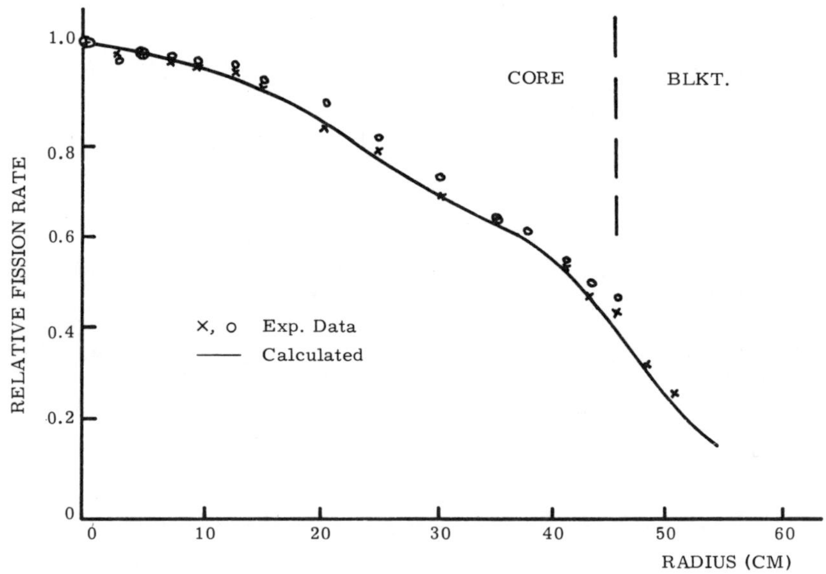


Figure 5.6 Enriched uranium fission counter radial traverse, Set No. 2, ZPR-III, Assembly 39.

TABLE 5.4 *Energy Modal Calculation, Central Spectral Indices — ZPR-III, Assembly 39*

| | Set No. 1 | Set No. 2 | $S_4$† | expt† |
|---|---|---|---|---|
| $\bar{\sigma}_f^{26}/\bar{\sigma}_f^{25}$ | 0.115 | 0.109 | 0.120 | 0.112 |
| $\bar{\sigma}_f^{28}/\bar{\sigma}_f^{25}$ | 0.049 | 0.046 | 0.052 | 0.047 |
| $\bar{\sigma}_f^{40}/\bar{\sigma}_f^{25}$ | 0.395 | 0.352 | 0.394 | 0.326 |
| $\bar{\sigma}_f^{24}/\bar{\sigma}_f^{25}$ | 0.400 | 0.350 | 0.383 | 0.347 |
| $\bar{\sigma}_f^{49}/\bar{\sigma}_f^{25}$ | 1.273 | 1.205 | 1.230 | 1.181 |
| $\bar{\sigma}_f^{33}/\bar{\sigma}_f^{25}$ | 1.597 | 1.563 | 1.569 | 1.523 |
| $\bar{\sigma}_c^{28}/\bar{\sigma}_f^{25}$ | 0.107 | 0.118 | 0.114 | |

† These are for ZPR-III, Assembly 31, which had the same composition as Assembly 39, and were taken from Reference 12.

calculations. From the fact that the $Pu^{239}$ fission ratios are overpredicted, it may be inferred that the relative flux below $10^5$ ev is somewhat underpredicted. In general, the modal calculations predict these relative reaction rates as well as the $S_4$ 16-group calculations. A commensurate ability to predict conversion ratios and relative isotopic depletion rates is implied.

Because the modal calculations predict $\bar{\sigma}_f^{28}/\bar{\sigma}_f^{25}$ well at the center of the core and predict the radial traverses well, it follows that $\bar{\sigma}_f^{28}/\bar{\sigma}_f^{25}$ is predicted well at all radial positions.

The preceding results do not imply that the modal diffusion approximation is as good as an $S_4$ calculation. Owing to the relatively large size of the core, a 16-group diffusion theory calculation would probably yield results comparable to those obtained with the $S_4$ calculation. However, these results do imply that the spatially dependent spectrum may be constructed from two modes for this simple assembly, and that relative reaction rates and the critical mass obtained from the modal approximation agree with experiment as well as those obtained from the multigroup finite difference calculation. The calculational effort involved in this modal approximation is equivalent to that in a 2-group calculation, which is orders of magnitude less than is required for a 16-group calculation.

## 5.3 Semidirect Method: Space Modes

The purposes of this section are to illustrate how the space modal formalism can be employed to account for the spatial dependence of the neutron flux, to demonstrate how conventional spatially independent multigroup theory can be adapted to solve the space modal equation, and to evaluate the efficacy of a low-order space modal calculation of certain neutronic properties of interest. To this end, the criticality, radial fission counter traverses, and central spectral indices have been calculated for ZPR-III, Assembly 39, which was described in the previous section.

When the neutron flux is represented by

$$\phi(r, E) = \psi_1(r)\rho_1(E) + \psi_2(r)\rho_2(E), \tag{5.13}$$

where the $\psi_i$ are known functions, the space modal equation which must be solved for the $\rho_i(E)$ follows from Equation 3.21,

$$\begin{aligned}\begin{bmatrix}\Sigma^{11}(E) & \Sigma^{12}(E)\\ \Sigma^{21}(E) & \Sigma^{22}(E)\end{bmatrix}\begin{bmatrix}\rho_1(E)\\ \rho_2(E)\end{bmatrix} &= \int_E^\infty dE'\begin{bmatrix}\Sigma_{et}^{11}(E'/E) & \Sigma_{et}^{12}(E'/E)\\ \Sigma_{et}^{21}(E'/E) & \Sigma_{et}^{22}(E'/E)\end{bmatrix}\begin{bmatrix}\rho_1(E')\\ \rho_2(E')\end{bmatrix}\\ &+ \int_E^\infty dE'\begin{bmatrix}\Sigma_{in}^{11}(E'/E) & \Sigma_{in}^{12}(E'/E)\\ \Sigma_{in}^{21}(E'/E) & \Sigma_{in}^{22}(E'/E)\end{bmatrix}\begin{bmatrix}\rho_1(E')\\ \rho_2(E')\end{bmatrix}\\ &+ \chi(E)\int_0^\infty dE'\begin{bmatrix}\nu\Sigma_f{}^{11}(E') & \nu\Sigma_f{}^{12}(E')\\ \nu\Sigma_f{}^{21}(E') & \nu\Sigma_f{}^{22}(E')\end{bmatrix}\begin{bmatrix}\rho_1(E')\\ \rho_2(E')\end{bmatrix}.\end{aligned} \tag{5.14}$$

The elements of the matrices are bilinear spatial averages defined by Equations 3.22.

Equation 5.14 can be cast in multigroup format, as was discussed in Chapter 3, resulting in Equation 3.25 with $N = 2$. The generic form of the group equations is

$$\begin{aligned}\Sigma_R{}^{11}(n)\rho_1(n) + \Sigma_R{}^{12}(n)\rho_2(n) &= S^1(n),\\ \Sigma_R{}^{21}(n)\rho_1(n) + \Sigma_R{}^{22}(n)\rho_2(n) &= S^2(n), \qquad n = 1,\cdots, M,\end{aligned} \tag{5.15}$$

where

$$\begin{aligned}S^j(n) = \sum_{n'=n+1}^{M} &[(\Sigma_{et}^{j1}(n'/n) + \Sigma_{in}^{j1}(n'/n))\rho_1(n')\\ &+ (\Sigma_{et}^{j2}(n'/n) + \Sigma_{in}^{j2}(n'/n))\rho_2(n')]\\ &+ \chi(n)\sum_{n'=1}^{M}[\nu\Sigma_f{}^{j1}(n')\rho_1(n') + \nu\Sigma_f{}^{j2}(n')\rho_2(n')], \qquad j = 1, 2.\end{aligned} \tag{5.16}$$

The superscripts refer to the spatial mode and weighting function used to perform the bilinear average. The argument $n$ refers to the group, and $\Sigma_R{}^{ij}$, $\nu\Sigma_f{}^{ij}$, and $\Sigma^{ij}(n'/n)$ are standard multigroup removal, *nu*-fission, and transfer cross sections that have been bilinearly averaged over the volume of the reactor, and are defined by Equations 3.26. Thus, all cross-section quantities in Equations 5.15 and 5.16 can be obtained from standard multigroup cross-section sets.

The set of Equations 5.15 and 5.16 can be solved iteratively by first guessing $\rho_i(n)$ for all $i$ and $n$, calculating $S^j(n)$ from Equation 5.16, then evaluating $\rho_i(n)$ from Equation 5.15, etc. To prevent the iteration from degenerating, the $\rho_i(n)$ are normalized by multiplication by

$$\xi_P \equiv \left[ \frac{\sum_{n=1}^{M} \left(\rho_1(n) + \rho_2(n)\right)^2_{P-1 \text{ iteration}}}{\sum_{n=1}^{M} \left(\rho_1(n) + \rho_2(n)\right)^2_{P \text{ iteration}}} \right]^{1/2} \tag{5.17}$$

after each iteration. The iteration should be continued until either $\xi_P$ obtains a constant value, which is an eigenvalue iteration; or until each $\rho_i(n)$ converges to a constant value, which is a flux iteration; or until each $S^j(n)$ converges to a constant value, which is a source iteration. The effective multiplication factor is the reciprocal of the converged value of $\xi_P$. These iterative procedures are analogous to techniques standardly employed in solving the conventional spatially independent multigroup equations.[10]

The spatial modes $\psi_1$ and $\psi_2$ are determined from the experimental fission counter traverses plotted in Figures 5.3 through 5.6. Mode $\psi_1$ is taken as the depleted fission counter traverse, and $\psi_2$ is taken as the difference between the enriched and depleted fission counter traverses. The weighting functions are the modes; i.e., $\alpha_1 = \psi_1$ and $\alpha_2 = \psi_2$. The bilinear averaging specified by Equations 3.22 is performed numerically, using Simpson's rule and approximating the spatial derivative operator by the central difference operator. The depleted fission counter traverse is a measure of the spatial distribution of the neutron flux above $\sim 1$ Mev and the enriched fission counter traverse is a measure of the over-all neutron flux distribution. Since the depleted and enriched fission counter traverses are similar, the second mode is chosen as the difference between the two traverses in order to ensure that the two equations for the $\rho_i$ are linearly independent.

Experimental data are used in constructing spatial modes principally to illustrate another means of obtaining modes. The energy modal calculation of the previous section could be used for this purpose with substantially the same results, since the predicted fission counter traverses are in good agreement with the experimental data.

Equations 5.15 and 5.16 are solved iteratively, using an eigenvalue convergence criterion of 0.002. The 16-group ANL-635 cross-section set is used. The modal combining coefficients thus obtained, $\rho_1$ and $\rho_2$, are listed in Table 5.5. Mode 1 is dominant, except for low energy neutrons. This is consistent with the interpretation of the depleted fission counter traverse (mode 1) as the distribution of the neutron flux above ~1 Mev, and of the difference between the enriched and depleted fission counter traverses (mode 2) as being due to lower energy neutrons.

TABLE 5.5 *Modal Combining Coefficients, Space Modal Calculation —ZPR-III, Assembly* 39

| $\Delta E$(Mev) | $\int_{\Delta E} \rho_1(E)\, dE$ | $\int_{\Delta E} \rho_2(E)\, dE$ |
|---|---|---|
| ∞–3.668 | 0.033 | — |
| 3.668–2.225 | 0.067 | — |
| 2.225–1.35 | 0.116 | 0.002 |
| 1.35–0.825 | 0.186 | 0.028 |
| 0.825–0.5 | 0.269 | 0.086 |
| 0.5–0.3 | 0.281 | 0.114 |
| 0.3–0.18 | 0.219 | 0.095 |
| 0.18–0.11 | 0.150 | 0.074 |
| 0.11–0.067 | 0.105 | 0.034 |
| 0.067–0.0407 | 0.074 | 0.046 |
| 0.0407–0.025 | 0.025 | 0.014 |
| 0.025–0.015 | 0.018 | 0.010 |
| 0.015–0.0091 | 0.003 | 0.003 |
| 0.0091–0.0055 | 0.001 | 0.001 |
| 0.0055–0.0021 | — | 0.001 |
| 0.0021–0.0005 | — | — |

Radial fission counter traverses calculated with Equation 5.13 compare favorably with the experimental data plotted in Figure 5.7. It should be noted in this respect that, although the spatial modes were constructed from the experimental fission counter traverses, it is necessary that the modal combining coefficients be properly calculated in order to obtain the correct fission counter traverses. This agreement may be considered as a check on the internal consistency of the calculation.

An effective multiplication factor, prescribed as the reciprocal of the asymptotic value of Equation 5.17, of 1.018 was calculated. From the measured reactivity worth of $U^{235}$ at the edge of the core,[3] it can be determined that $\Delta M/M = 4.3\ \Delta k/k$. Thus, the space modal calculation underpredicts the critical mass by approximately 7.7%. If the effects of boundary irregularity, gaps, and heterogeneities are accounted for, the

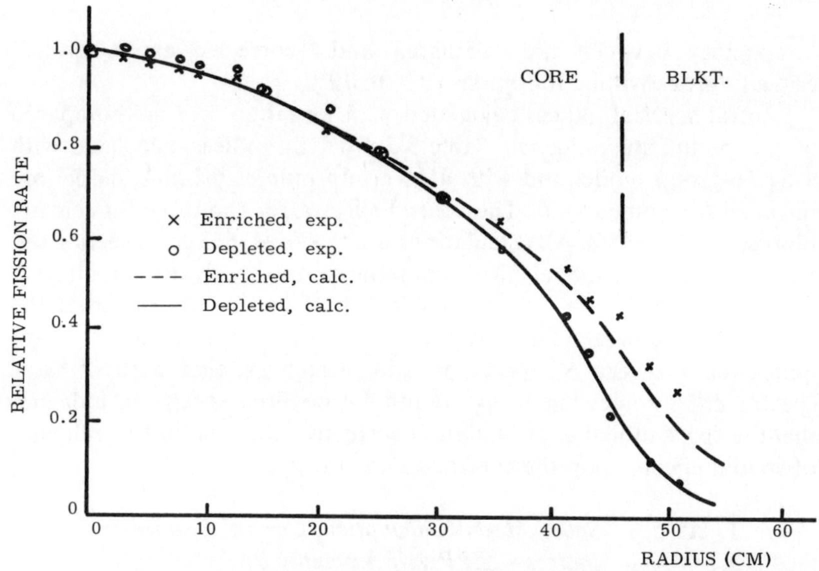


FIGURE 5.7: Uranium Fission Counter Radial Traverses Space Modal Calculation, ZPR-III, Assy. 39

Figure 5.7 Uranium fission counter radial traverses, space modal calculation, ZPR-III, Assembly 39.

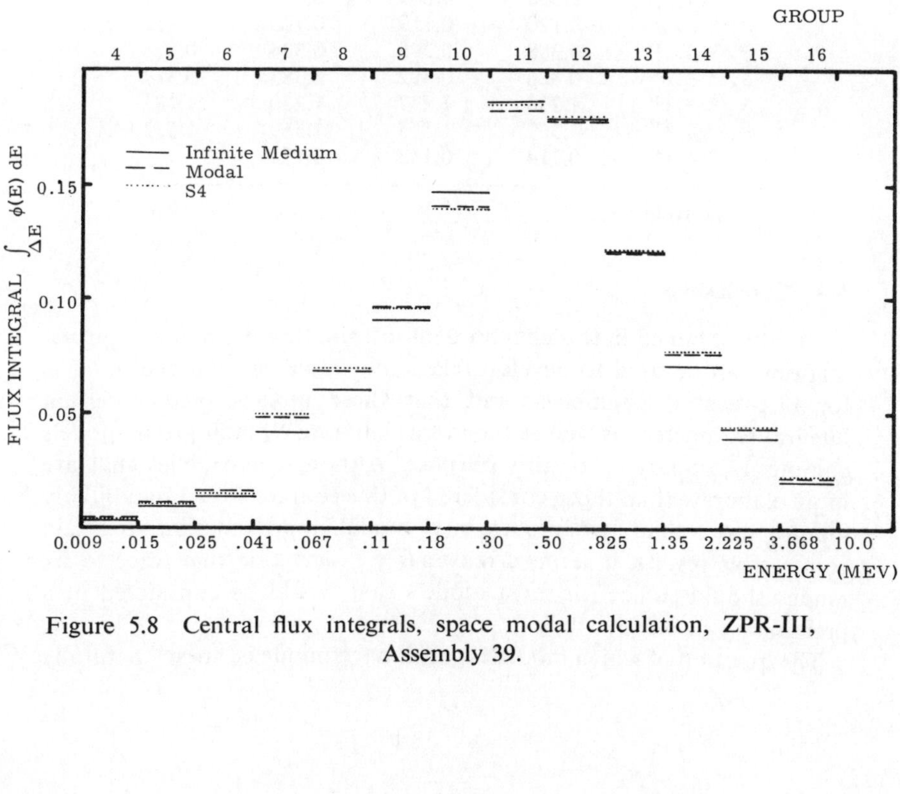


Figure 5.8 Central flux integrals, space modal calculation, ZPR-III, Assembly 39.

discrepancy between the calculated and "corrected experimental" critical masses will be more like 10% to 12%.

Central spectral indices calculated with Equation 5.13 are compared with experimental values in Table 5.6. Spectral indices calculated with a $S_4$ 16-group model and with a 16-group infinite medium model are included for comparison. The central spectra for the three models are plotted in Figure 5.8. All calculations use the ANL-635 cross-section set.

The influence of spatial effects upon the central spectrum is small, as is apparent from Figure 5.8 and Table 5.6. However, the fact that the central spectrum calculated by the space modal approximation agrees quite well with the $S_4$ spectrum, and, moreover, that both of these spectra differ somewhat from the infinite medium spectrum, indicates that the space modal approximation correctly accounts for the influence of spatial effects upon the central spectrum.

TABLE 5.6 *Space Modal Calculation, Central Spectral Indices—ZPR-III, Assembly* 39

| | Space Modal | 16-Group ∞-Medium | 16-Group† $S_4$ | exp.† |
|---|---|---|---|---|
| $\bar{\sigma}_f^{28}/\bar{\sigma}_f^{25}$ | 0.050 | 0.049 | 0.052 | 0.047 |
| $\bar{\sigma}_f^{26}/\bar{\sigma}_f^{25}$ | 0.120 | 0.119 | 0.120 | 0.112 |
| $\bar{\sigma}_f^{40}/\bar{\sigma}_f^{25}$ | 0.388 | 0.393 | 0.394 | 0.326 |
| $\bar{\sigma}_f^{24}/\bar{\sigma}_f^{25}$ | 0.386 | 0.392 | 0.383 | 0.347 |
| $\bar{\sigma}_f^{49}/\bar{\sigma}_f^{25}$ | 1.230 | 1.237 | 1.230 | 1.181 |
| $\bar{\sigma}_f^{23}/\bar{\sigma}_f^{25}$ | 1.569 | 1.575 | 1.569 | 1.523 |
| $\bar{\sigma}_c^{28}/\bar{\sigma}_f^{25}$ | 0.114 | 0.113 | 0.114 | |

† See Reference 12.

## 5.4 Conclusions

Results obtained in this chapter demonstrate that the modal approximations can be used to develop relatively simple calculational models for fast-reactor assemblies, and that these models predict certain integral parameters as well as the more elaborate $S_4$ multigroup models commonly employed for this purpose. Although assemblies that are more elaborate than those considered in this chapter would indubitably require more than two modes in a modal approximation to attain comparable results, it seems a reasonable expectation that three to six modes should suffice for most models that would be considered in a preliminary or survey study.

The question of which modal expansion technique is "best" naturally arises. This question can be discussed in very general terms. A modal approximation should take advantage of whatever knowledge of the solution is available, without unduly biasing the other features of the solution. As an example, energy modes could be constructed for ZPR-III, Assembly 39 from physical arguments; the energy modal approximation seemed natural. On the other hand, it would have been more difficult to obtain reasonable space modes without prior knowledge.

The direct method was appropriate for GODIVA; even if the spatial dependence had not been eliminated by the artifice of a leakage cross section, it would have been reasonable to use the fission and inelastic spectra and a fundamental mode spatial distribution as space-energy modes. However, use of the direct method for ZPR-III, Assembly 39, would require an estimate of the spatial dependence associated with each of the energy modes, or an estimate of the energy dependence associated with each of the space modes. The semidirect methods require less knowledge of the solution and are more flexible than the direct method.

# 6. Conclusions and Recommendations

The original purposes of this research were to codify the modal expansion method and to apply this method to obtain simple approximate calculational models for fast-reactor assemblies. In the course of attaining the second goal, it became apparent that simple analytical expressions for the spectrum and importance function in spatially independent fast-reactor assemblies were desirable. These three areas provide convenient classifications for an appraisal of the results of this research and for recommendation for future extensions.

## 6.1 Modal Expansion Method

A general formalism was outlined for obtaining an approximate set of equations with a reduced number of independent variables. The mathematical significance of the modal expansion, in terms of representation theory, was examined. Criteria were developed for choosing weighting functions that minimize the error in the modal expansion solution, in some sense. Many well-known approximation methods were shown to be special cases of the general modal expansion method.

Two areas in which further work may yield profitable results are the following:

1. A rigorous analysis of the modal expansion procedure by the methods of functional analysis. This may suggest criteria for choosing modes and weighting functions, and will certainly provide insight into the mathematical aspects of the procedure.

2. An examination of existing approximation schemes that may be classified as special cases of modal expansion methods. Such an examination would, possibly, clarify the nature of the approximation that was involved in terms of the error criteria and might suggest refinements or alternatives.

## 6.2 Applications to Fast-Reactor Calculations

The emphasis in this monograph is upon the application of modal expansion methods to obtain simple calculational models that yield approximate results. Such models were developed and successfully applied to evaluate reaction rates, critical mass, and spatial fission traverses for simple assemblies. An extension to more complicated assemblies is conceptually straightforward; however, the computational effort involved would require the use of a computer. Thus, there is a need to develop digital computer programs capable of handling a large number of modes. Standard techniques can be adapted, as was pointed out in Chapters 3 and 5.

The space modal (Section 3.2.2) and the energy modal (Section 3.2.3) approximations may be used as alternate steps in an iterative solution. Investigation of this possibility may lead to useful techniques for solving problems in which no reasonable estimate of modes is available or necessary.

There is no *a priori* reason that the use of the modal expansion method should be limited to developing simple approximate calculational models. If enough modes are employed, it is plausible that models may be developed which yield results to any desired accuracy, subject, naturally, to the uncertainty in nuclear data.

In particular, it is felt that the energy modal approximation offers an attractive and computationally economic alternative to the multigroup approximation, in that a given accuracy in the representation of the spectrum may be obtained with considerably fewer continuous energy modes than groups. Some qualitative arguments supporting this conjecture were offered in Section 3.4. This may have significant ramifications with respect to the calculation of such spectrum sensitive quantities as the Doppler coefficient, and with respect to the evaluation of reaction rates and reactivity coefficients, in general. The use of conventional approximations ($P_n$, $S_n$, moments, eigenfunction expansion, etc.) in conjunction with the energy modal transport approximation (Section 3.1) may result in powerful techniques that are more computationally economical than the standard techniques employing the multigroup approximation.

In the development of accurate calculational models, the mathematical criteria for the selection of modes and weighting operators, particularly the linear independence requirement, will be more important than the physically motivated criteria that were exploited in Chapter 5.

### 6.3 Analytical Formulas for Spectrum and Importance Function

It was demonstrated that the analytical formulas for the spatially independent spectrum and importance function, which were developed in Chapter 4, could be used to obtain approximate evaluations of integral parameters of interest in fast-reactor calculations. However, the choice of the inelastic scattering parameter $T$ can be specified to within only a certain range of values by theoretical and experimental considerations. Because the statistical model treatment of inelastic scattering is not theoretically sound in the region below 2 Mev, and because the inelastic spectrum was assumed to be independent of the incident neutron energy, it is improbable that further theoretical work will resolve the ambiguity in the choice of $T$. A practical solution to this problem is to regard $T$ as a parameter to be determined by fitting the analytical spectrum to experiment. This fit could be made by requiring agreement of integral parameters or by requiring the calculated and experimental spectra, when the latter become available, to agree in some sense.

The treatment of the elastic scattering by defining effective probabilities, which were evaluated from continuous slowing down theory, is another area in which improvements may be made. One possibility is to use an alternate to continuous slowing down theory in evaluating the effective probabilities. Another approach, which might be promising, is to attempt to represent the elastic scattering kernel as a degenerate kernel, in which case the problem can be solved exactly. The neutron balance equation, Equation 4.1, may be written

$$F(E) = \chi(E) + \int_0^{E_0} K(E, E')F(E')\,dE', \tag{6.1}$$

where

$$K(E, E') = \frac{r_{et}(E')}{E'(1-\alpha)}\{-U(E' - E/\alpha) + U(E' - E)\} + \sum_{m=1}^{M} P^m(E) r_{in}^m(E'), \tag{6.2}$$

$U$ is the step function, and $E_0$ is some energy 10 Mev $< E_0 < \infty$. If it is practical to write

$$-U(E' - E/\alpha) + U(E' - E) = \sum_{i=1}^{N} \phi(E)\psi(E'), \tag{6.3}$$

then the kernel is degenerate,

$$K(E, E') = \sum_{i=1}^{N} \phi(E)\left[\frac{\psi(E')r_{et}(E')}{E'(1-\alpha)}\right] + \sum_{m=1}^{M} P^m(E)r_{in}^m(E'), \qquad (6.4)$$

and Equation 6.1 can be solved by standard techniques.[46]

The method of successive approximations might be applied to obtain expressions for the spectrum and importance function in intermediate reactors, also. The general solutions, Equations 4.6 and 4.22, are valid for the energy range above approximately 1 ev, below which "up-scattering" becomes important and the simple elastic scattering kernel is no longer valid. The approximation of the iterated kernels by a two-term series expansion in powers of $(1 - \alpha)$ is certainly more justified in fast than intermediate assemblies, but is probably satisfactory for the latter.

Use of effective scattering probabilities may distort the fine structure of the neutron spectrum. In fast reactors, the important fission and absorption cross sections are smoothly varying over much of the energy range, and the reaction rates that determine the integral properties are relatively insensitive to the fine structure of the spectrum.[18] For intermediate reactors, the important fission and absorption cross sections exhibit a pronounced resonance structure over at least part of the range of interest, and integral properties are indubitably more sensitive to the fine structure. This might limit the effectiveness of the analytical expressions in intermediate-reactor applications.

# Appendix A. Representation of a Linear Operator Equation on a Linear Vector Space

The modal expansion method is equivalent to representing a linear operator equation on a linear vector space that has the modes as bases. The theory which will be outlined is slightly different than that found in standard texts[21] in that it is only a partial representation which is sought. A multivariate equation (independent variables of $x_1, \cdots, x_n$) is represented on a vector space of only one of the variables ($x_1$). Examination of this representation leads to some insight regarding the choice of modes and weighting functions for the modal expansion method, as indicated in Chapter 2.

First, some definitions of linear vector space theory are reviewed.[41] A linear vector space is composed of a set of $N$ linearly independent vectors

$$|\psi_i(x_1)\rangle, \qquad i = 1, \cdots, N. \tag{A.1}$$

This set of vectors will be denoted by $\Sigma_N$. The number $N$ may be infinite and the distribution of vectors may be continuous, in general. Associated with each vector in $\Sigma_N$ is a conjugate vector

$$\langle \alpha_i(x_1)|, \qquad i = 1, \cdots, N, \tag{A.2}$$

of the dual space $\Sigma_N{}^*$ associated with $\Sigma_N$. The notation of Dirac[16] is used.

The scalar product of two vectors in $\Sigma_N$ is defined as the number

$$\langle u(x_1)|v(x_1)\rangle = \overline{\langle v(x_1)|u(x_1)\rangle}. \tag{A.3}$$

Because the $|\psi_i\rangle$ are linearly independent, they may be chosen orthonormal in the sense that the scalar product $\langle\alpha_i|$ upon $|\psi_j\rangle$ is

$$\langle\alpha_i(x_1)|\psi_j(x_1)\rangle = \delta_{ij}. \tag{A.4}$$

Completeness of the set $|\psi_i\rangle$ implies that any vector in $\Sigma_N$ may be represented in the form

$$|u(x_1)\rangle = \sum_{i=1}^{N} C_i|\psi_i(x_1)\rangle. \tag{A.5}$$

By making use of Equation A.4, this becomes

$$|u(x_1)\rangle = \sum_{i=1}^{N} |\psi_i(x_1)\rangle\langle\alpha_i(x_1')|u(x_1')\rangle, \tag{A.6}$$

which leads to the closure relation commonly used to denote completeness

$$1 \equiv \sum_{i=1}^{N} |\psi_i(x_1)\rangle\langle\alpha_i(x_1')|. \tag{A.7}$$

A linear operator equation

$$H(x_1,\cdots, x_n)|\phi(x_1,\cdots, x_n)\rangle - |f(x_1,\cdots, x_n)\rangle = 0 \tag{A.8}$$

may be represented on $\Sigma_N$ by formally inserting the identity Equation A.7 into Equation A.8.

$$\sum_{ij=1}^{N} |\psi_i(x_1)\rangle\langle\alpha_i(x_1')|H(x_1',\cdots, x_n)|\psi_j(x_1')\rangle\langle\alpha_j(x_1'')|\phi(x_1'',\cdots, x_n)\rangle$$
$$- \sum_{i=1}^{N} |\psi_i(x_1)\rangle\langle\alpha_i(x_1')|f(x_1',\cdots, x_n)\rangle = 0. \tag{A.9}$$

Equation A.9 is exact in that its solution $\langle\alpha_j|\phi\rangle$ may be used to construct an exact solution of Equation A.8 of the form of Equation A.6.

# Appendix B. Two-Energy Mode–Two-Region Problem

As indicated in Chapter 3, it is possible to obtain analytical solutions to Equation 3.30 for certain simple geometries. In this appendix, analytic solutions are developed for the two-energy mode equations in two-region geometries. Only those geometries for which the neutron flux depends upon a single spatial variable are considered: infinite slab, infinite cylinder, and sphere. Each region is assumed to be uniform and homogeneous. In each region, an equation of the form of Equation 3.30 may be written:

$$\begin{bmatrix} (-D^{11}\nabla^2 + \Sigma^{11}) & (-D^{12}\nabla^2 + \Sigma^{12}) \\ (-D^{21}\nabla^2 + \Sigma^{21}) & (-D^{22}\nabla^2 + \Sigma^{22}) \end{bmatrix} \begin{bmatrix} \rho_1(r) \\ \rho_2(r) \end{bmatrix} = \begin{bmatrix} f^1(r) \\ f^2(r) \end{bmatrix}. \tag{B.1}$$

The interface of the two regions is located at $r = R_0$, the extrapolated boundary is at $r = \tilde{a}$, and the reactor is symmetrical about $r = 0$. The interface boundary conditions, Equations 3.32, are

$$\begin{bmatrix} \rho_1(R_0) \\ \rho_2(R_0) \end{bmatrix}_1 = \begin{bmatrix} \rho_1(R_0) \\ \rho_2(R_0) \end{bmatrix}_2, \tag{B.2a}$$

$$\begin{bmatrix} D^{11} & D^{12} \\ D^{21} & D^{22} \end{bmatrix} \begin{bmatrix} \nabla\rho_1(R_0) \\ \nabla\rho_2(R_0) \end{bmatrix}\Bigg|_1 = \begin{bmatrix} D^{11} & D^{12} \\ D^{21} & D^{22} \end{bmatrix} \begin{bmatrix} \nabla\rho_1(R_0) \\ \nabla\rho_2(R_0) \end{bmatrix}\Bigg|_2. \tag{B.2b}$$

The condition that the flux vanish at the extrapolated boundary $\tilde{a}$ is

$$\begin{bmatrix} \rho_1(\tilde{a}) \\ \rho_2(\tilde{a}) \end{bmatrix}\Bigg|_2 = 0. \tag{B.3}$$

If the source is symmetric about $r = 0$, the remaining conditions are that the flux be symmetric about $r = 0$, and be everywhere finite.

Due to the formal analogy between Equation B.1 and the conventional two-group equations, a solution to Equation B.1 may be developed along lines similar to those followed in obtaining a solution to the two-group, two-region problem.[39] Homogeneous solutions of Equation B.1 are sought that satisfy

$$\begin{aligned} \nabla^2\tilde{U}(r) + \alpha_+{}^2\tilde{U}(r) &= 0, \\ \nabla^2\tilde{V}(r) + \alpha_-{}^2\tilde{V}(r) &= 0, \end{aligned} \tag{B.4}$$

in region 1, and in region 2, satisfy

$$\begin{aligned} \nabla^2\tilde{W}(r) + \beta_+{}^2\tilde{W}(r) &= 0, \\ \nabla^2\tilde{Z}(r) + \beta_-{}^2\tilde{Z}(r) &= 0. \end{aligned} \tag{B.5}$$

Depending upon whether $\alpha_+{}^2$ is positive, negative, or complex, the $\tilde{U}(r)$ involves trigonometric functions (Bessel functions with real argument), hyperbolic functions (Bessel functions with imaginary argument), or a combination of both, respectively. Assuming, for the moment, that $\alpha_+{}^2$ and $\beta_+{}^2 > 0$, $\alpha_-{}^2$ and $\beta_-{}^2 < 0$, and that all four are real, the homogeneous solutions to Equations B.4 and B.5 are summarized as follows:

| | Slab | Sphere | Cylinder |
|---|---|---|---|
| $\tilde{U}$ | $\sin \alpha_+ r, \cos \alpha_+ r$ | $\dfrac{\sin \alpha_+ r}{r}, \dfrac{\cos \alpha_+ r}{r}$ | $J_0(\alpha_+ r), Y_0(\alpha_+ r)$ |
| $\tilde{V}$ | $\sinh \alpha_- r, \cosh \alpha_- r$ | $\dfrac{\sinh \alpha_- r}{r}, \dfrac{\cosh \alpha_- r}{r}$ | $I_0(\alpha_- r), K_0(\alpha_- r)$ |
| $\tilde{W}$ | $\sin \beta_+ r, \cos \beta_+ r$ | $\dfrac{\sin \beta_+ r}{r}, \dfrac{\cos \beta_+ r}{r}$ | $J_0(\beta_+ r), Y_0(\beta_+ r)$ |
| $\tilde{Z}$ | $\sinh \beta_- r, \cosh \beta_- r$ | $\dfrac{\sinh \beta_- r}{r}, \dfrac{\cosh \beta_- r}{r}$ | $I_0(\beta_- r), K_0(\beta_- r)$ |

Specializing to a source that is uniform within each region, particular solutions of Equation B.1 are

$$\rho_{1p}(r) = \frac{\Sigma^{22}f^1 - \Sigma^{12}f^2}{\Sigma^{11}\Sigma^{22} - \Sigma^{21}\Sigma^{12}} \equiv S^1, \qquad \rho_{2p}(r) = \frac{\Sigma^{11}f^2 - \Sigma^{21}f^1}{\Sigma^{11}\Sigma^{22} - \Sigma^{21}\Sigma^{12}} \equiv S^2 \tag{B.7}$$

for each region. The homogeneous solutions in Region 1 are

$$\begin{aligned}\rho_{1h}(r) &= A_1\tilde{U}(r) + C_1\tilde{V}(r),\\ \rho_{2h}(r) &= A_2\tilde{U}(r) + C_2\tilde{V}(r),\end{aligned} \tag{B.8}$$

and in region 2 are,

$$\begin{aligned}\rho_{1h}(r) &= E_1\tilde{W}(r) + H_1\tilde{Z}(r),\\ \rho_{2h}(r) &= E_2\tilde{W}(r) + H_2\tilde{Z}(r).\end{aligned} \tag{B.9}$$

The constants $A$, $C$, $E$, and $H$ are not all independent but must be related in such a way that Equations B.8 and B.9 satisfy the homogeneous form of Equation B.1. At this point, it is convenient to introduce the subscript notation $c$ and $r$ to refer to region 1 (core) and region 2 (reflector). When Equations B.8 are substituted into the homogeneous form of Equation B.1, the following relations are obtained:

$$\begin{aligned}0 &= [(D_c^{11}\alpha_+^2 + \Sigma_c^{11})A_1 + (D_c^{12}\alpha_+^2 + \Sigma_c^{12})A_2]\tilde{U}(r)\\ &\quad + [(D_c^{11}\alpha_-^2 + \Sigma_c^{11})C_1 + (D_c^{12}\alpha_-^2 + \Sigma_c^{12})C_2]\tilde{V}(r),\\ 0 &= [(D_c^{21}\alpha_+^2 + \Sigma_c^{21})A_1 + (D_c^{22}\alpha_+^2 + \Sigma_c^{22})A_2]\tilde{U}(r)\\ &\quad + [(D_c^{21}\alpha_-^2 + \Sigma_c^{21})C_1 + (D_c^{22}\alpha_-^2 + \Sigma_c^{22})C_2]\tilde{V}(r).\end{aligned} \tag{B.10}$$

Because $\tilde{U}(r)$ and $\tilde{V}(r)$ are linearly independent, Equations B.10 can be satisfied identically only if the terms in square brackets vanish. This yields two relations between $A_1$ and $A_2$, and two relations between $C_1$ and $C_2$.

$$\frac{A_2}{A_1} = -\frac{(D_c^{11}\alpha_+^2 + \Sigma_c^{11})}{(D_c^{12}\alpha_+^2 + \Sigma_c^{12})} = -\frac{(D_c^{21}\alpha_+^2 + \Sigma_c^{21})}{(D_c^{22}\alpha_+^2 + \Sigma_c^{22})} \equiv a_1, \tag{B.11}$$

$$\frac{C_2}{C_1} = -\frac{(D_c^{11}\alpha_-^2 + \Sigma_c^{11})}{(D_c^{12}\alpha_-^2 + \Sigma_c^{12})} = -\frac{(D_c^{21}\alpha_-^2 + \Sigma_c^{21})}{(D_c^{22}\alpha_-^2 + \Sigma_c^{22})} \equiv a_2. \tag{B.12}$$

Equations B.11 and B.12 can be satisfied only for certain values of $\alpha_+^2$ and $\alpha_-^2$. Solving these equations yields

$$\alpha_\pm^2 = -\frac{(\Sigma_c^{11}D_c^{22} + \Sigma_c^{22}D_c^{11} - \Sigma_c^{12}D_c^{21} - \Sigma_c^{21}D_c^{12})}{2(D_c^{11}D_c^{22} - D_c^{21}D_c^{12})} \times \left[1 \pm \sqrt{1 + \frac{4(D_c^{21}D_c^{12} - D_c^{11}D_c^{22})(\Sigma_c^{11}\Sigma_c^{22} - \Sigma_c^{12}\Sigma_c^{21})}{(\Sigma_c^{11}D_c^{22} + \Sigma_c^{22}D_c^{11} - \Sigma_c^{12}D_c^{21} - \Sigma_c^{21}D_c^{12})^2}}\right] \tag{B.13}$$

TABLE B.1 *General Solutions*

| | SLAB | SPHERE | CYLINDER |
|---|---|---|---|
| $\rho_{1c}(r)^a$ | $A_1 \cos \alpha_+ r +$ | $A_1 \frac{\sin \alpha_+ r}{r} +$ | $A_1 J_o(\alpha_+ r) +$ |
| | $C_1 \cosh \alpha_- r + S_c^1$ | $C_1 \frac{\sinh \alpha_- r}{r} + S_c^1$ | $C_1 I_o(\alpha_- r) + S_c^1$ |
| $\rho_{2c}(r)^a$ | $a_1 A_1 \cos \alpha_+ r +$ | $a_1 A_1 \frac{\sin \alpha_+ r}{r} +$ | $a_1 A_1 J_o(\alpha_+ r) +$ |
| | $a_2 C_1 \cosh \alpha_- r + S_c^2$ | $a_2 C_1 \frac{\sinh \alpha_- r}{r} + S_c^2$ | $a_2 C_1 I_o(\alpha_- r) + S_c^2$ |
| $\rho_{1r}(r)^a$ | $D_1 \sin \beta_+(\tilde{a}-r) +$ | $D_1 \frac{\sin \beta_+(\tilde{a}-r)}{r} +$ | $E_1 \left[ Y_o(\beta_+ r) - \frac{Y_o(\beta_+ \tilde{a})}{J_o(\beta_+ \tilde{a})} J_o(\beta_+ r) \right] +$ |
| | $H_1 \sinh \beta_-(\tilde{a}-r) + x_1(r)^b$ | $H_1 \frac{\sinh \beta_-(\tilde{a}-r)}{r} + x_1(r)$ | $H_1 \left[ K_o(\beta_- r) - \frac{K_o(\beta_- \tilde{a})}{I_o(\beta_- \tilde{a})} I_o(\beta_- r) \right] + x_1(r)$ |
| $\rho_{2r}{}^a$ | $a_3 D_1 \sin \beta_+(\tilde{a}-r) +$ | $a_3 D_1 \frac{\sin \beta_+(\tilde{a}-r)}{r} +$ | $a_3 E_1 \left[ Y_o(\beta_+ r) - \frac{Y_o(\beta_+ \tilde{a})}{J_o(\beta_+ \tilde{a})} J_o(\beta_+ r) \right] +$ |
| | $a_4 H_1 \sinh \beta_-(\tilde{a}-r) + x_2(r)$ | $a_4 D_1 \frac{\sinh \beta_-(\tilde{a}-r)}{r} + x_2(r)$ | $a_4 H_1 \left[ K_o(\beta_- r) - \frac{K_o(\beta_- \tilde{a})}{I_o(\beta_- \tilde{a})} I_o(\beta_- r) \right] + x_2(r)$ |

[a] The subscripts c and r refer to region 1 (core) and region 2 (reflector).

[b] $x_1(r)$ and $x_2(r)$ are defined in Table B.2.

TABLE B.2 $(x)r$

| | $x_1$ | $x_2$ |
|---|---|---|
| SLAB | $S_r^1 + \left[\dfrac{S_r^2 - a_4 S_r^1}{(a_4 - a_3)\sin\beta_+\tilde{a}}\right]\sin\beta_+ r + \left[\dfrac{S_r^2 - a_3 S_r^1}{(a_3 - a_4)\sinh\beta_-\tilde{a}}\right]\sinh\beta_- r$ | $S_r^2 + a_3\left[\dfrac{S_r^2 - a_4 S_r^1}{(a_4 - a_3)\sin\beta_+\tilde{a}}\right]\sin\beta_+ r + a_4\left[\dfrac{S_r^2 - a_3 S_r^1}{(a_3 - a_4)\sinh\beta_-\tilde{a}}\right]\sinh\beta_- r$ |
| SPHERE | $S_r^1 + \left[\dfrac{(S_r^2 - a_4 S_r^1)\tilde{a}}{(a_4 - a_3)\sin\beta_+\tilde{a}}\right]\dfrac{\sin\beta_+ r}{r} + \left[\dfrac{(S_r^2 - a_3 S_r^1)\tilde{a}}{(a_3 - a_4)\sinh\beta_-\tilde{a}}\right]\dfrac{\sinh\beta_- r}{r}$ | $S_r^2 + a_3\left[\dfrac{(S_r^2 - a_4 S_r^1)\tilde{a}}{(a_4 - a_3)\sin\beta_+\tilde{a}}\right]\dfrac{\sin\beta_+ r}{r} + a_4\left[\dfrac{(S_r^2 - a_3 S_r^1)\tilde{a}}{(a_3 - a_4)\sinh\beta_-\tilde{a}}\right]\dfrac{\sinh\beta_- r}{r}$ |
| CYLINDER | $S_r^1 + \left[\dfrac{S_r^2 - a_4 S_r^1}{(a_4 - a_3)J_o(\beta_+\tilde{a})}\right]J_o(\beta_+ r) + \left[\dfrac{S_r^2 - a_3 S_r^1}{(a_3 - a_4)I_o(\beta_-\tilde{a})}\right]I_o(\beta_- r)$ | $S_r^2 + a_3\left[\dfrac{S_r^2 - a_4 S_r^1}{(a_4 - a_3)J_o(\beta_+\tilde{a})}\right]J_o(\beta_+ r) + a_4\left[\dfrac{S_r^2 - a_3 S_r^1}{(a_3 - a_4)I_o(\beta_-\tilde{a})}\right]I_o(\beta_- r)$ |

TABLE B.3 *Certain Functions*

| | W | W′ |
|---|---|---|
| Slab | $\sin \beta_+(\tilde{a}-R_o)$ | $-\beta_+ \cos \beta_+(\tilde{a}-R_o)$ |
| Sphere | $\dfrac{\sin \beta_+(\tilde{a}-R_o)}{R_o}$ | $-\dfrac{1}{R_o}\left(\beta_+ \cos \beta_+(\tilde{a}-R_o) + \dfrac{\sin \beta_+(\tilde{a}-R_o)}{R_o}\right)$ |
| Cylinder | $Y_o(\beta_+R_o) - \dfrac{Y_o(\beta_+\tilde{a})}{J_o(\beta_+\tilde{a})} J_o(\beta_+R_o)$ | $-\beta_+\left(Y_1(\beta_+R_o) - \dfrac{Y_o(\beta_+\tilde{a})}{J_o(\beta_+\tilde{a})} J_1(\beta_+R_o)\right)$ |

| | Z | Z′ |
|---|---|---|
| Slab | $\sinh \beta_-(\tilde{a}-R_o)$ | $-\beta_- \cosh \beta_-(\tilde{a}-R_o)$ |
| Sphere | $\dfrac{\sinh \beta_-(\tilde{a}-R_o)}{R_o}$ | $-\dfrac{1}{R_o}\left(\beta_- \cosh \beta_-(\tilde{a}-R_o) + \dfrac{\sinh \beta_-(\tilde{a}-R_o)}{R_o}\right)$ |
| Cylinder | $K_o(\beta_-R_o) - \dfrac{K_o(\beta_-\tilde{a})}{I_o(\beta_-\tilde{a})} I_o(\beta_-R_o)$ | $-\beta_-\left(K_1(\beta_-R_o) + \dfrac{K_o(\beta_-\tilde{a})}{I_o(\beta_-\tilde{a})} I_1(\beta_-R_o)\right)$ |

TABLE B.3 *(continued)*

| | U | U′ |
|---|---|---|
| Slab | $\cos \alpha_+ R_o$ | $-\alpha_+ \sin \alpha_+ R_o$ |
| Sphere | $\frac{\sin \alpha_+ R_o}{R_o}$ | $\frac{1}{R_o}\left(\alpha_+ \cos \alpha_+ R_o - \frac{\sin \alpha_+ R_o}{R_o}\right)$ |
| Cylinder | $J_o(\alpha_+ R_o)$ | $-\alpha_+ J_1(\alpha_+ R_o)$ |

| | V | V′ |
|---|---|---|
| Slab | $\cosh \alpha_- R_o$ | $\alpha_- \sinh \alpha_- R_o$ |
| Sphere | $\frac{\sinh \alpha_- R_o}{R_o}$ | $\frac{1}{R_o}\left(\alpha_- \cosh \alpha_- R_o - \frac{\sinh \alpha_- R_o}{R_o}\right)$ |
| Cylinder | $I_o(\alpha_- R_o)$ | $\alpha_- I_1(\alpha_- R_o)$ |

When Equations B.9 are substituted into the homogeneous form of Equation B.1, reasoning identical to that just used leads to

$$\frac{E_2}{E_1} = -\frac{D_r^{11}\beta_+^{\,2} + \Sigma_r^{11}}{D_r^{12}\beta_+^{\,2} + \Sigma_r^{12}} = -\frac{D_r^{21}\beta_+^{\,2} + \Sigma_r^{21}}{D_r^{22}\beta_+^{\,2} + \Sigma_r^{22}} \equiv a_3, \qquad \text{(B.14)}$$

$$\frac{H_2}{H_1} = -\frac{D_r^{11}\beta_-^{\,2} + \Sigma_r^{11}}{D_r^{12}\beta_-^{\,2} + \Sigma_r^{12}} = -\frac{D_r^{21}\beta_-^{\,2} + \Sigma_r^{21}}{D_r^{22}\beta_-^{\,2} + \Sigma_r^{22}} \equiv a_4, \qquad \text{(B.15)}$$

and

$$\beta_\pm^{\,2} = -\frac{\Sigma_r^{11}D_r^{22} + \Sigma_r^{22}D_r^{11} - \Sigma_r^{12}D_r^{21} - \Sigma_r^{21}D_r^{12}}{2(D_r^{11}D_r^{22} - D_r^{21}D_r^{12})} \times \left[1 \pm \sqrt{1 + \frac{4(D_r^{21}D_r^{12} - D_r^{11}D_r^{22})(\Sigma_r^{11}\Sigma_r^{22} - \Sigma_r^{12}\Sigma_r^{21})}{(\Sigma_r^{11}D_r^{22} + \Sigma_r^{22}D_r^{11} - \Sigma_r^{12}D_r^{21} - \Sigma_r^{21}D_r^{12})^2}}\,\right]. \qquad \text{(B.16)}$$

The most general solutions in each region are obtained as the sum of the homogeneous solution, given by Equations B.8 or B.9, and the particular solution, given in the case of a uniform source, by Equations B.7. When the extrapolated boundary condition, Equation B.3, and the requirement that the flux be finite and symmetric about $r = 0$ are applied, the solutions of Equation B.1 may be written as shown in Table B.1.

When the general solutions of Table B.1 are substituted into the interface boundary conditions specified by Equations B.2, a set of four simultaneous equations that must be solved for $A_1$, $C_1$, $D_1$, and $H_1$ are obtained. These four equations may be written as a matrix relation

$$\mathsf{P} \cdot \mathsf{C} = \mathsf{S}, \qquad \text{(B.17)}$$

where

$$\mathsf{P} \equiv \begin{bmatrix} U & V & -W & -Z \\ a_1U & a_2V & -a_3W & -a_4Z \\ (D_c^{11} + a_1D_c^{12})U' & (D_c^{11} + a_2D_c^{12})V' & -(D_r^{11} + a_3D_r^{12})W' & -(D_r^{11} + a_4D_r^{12})Z' \\ (D_c^{21} + a_1D_c^{22})U' & (D_c^{21} + a_2D_c^{22})V' & -(D_r^{21} + a_3D_r^{22})W' & -(D_r^{21} + a_4D_r^{22})Z' \end{bmatrix}$$

$$\mathsf{C} \equiv \begin{bmatrix} A_1 \\ C_1 \\ D_1 \\ H_1 \end{bmatrix}, \quad \mathsf{S} \equiv \begin{bmatrix} x_1(R_0) - S_c^{1} \\ x_2(R_0) - S_c^{2} \\ D_r^{11}x_1'(R_0) + D_r^{12}x_2'(R_0) \\ D_r^{21}x_1'(R_0) + D_r^{22}x_2'(R_0) \end{bmatrix}. \qquad \text{(B.18)}$$

Here, $x_1'$ and $x_2'$ are derivatives of the quantities $x_1$ and $x_2$ defined in Table B.2, while $U$, $V$, $W$, $Z$, $U'$, $V'$, $W'$, and $Z'$ are defined in Table B.3.

To repeat, the general solutions are given by Table B.1, with the coefficients determined from

$$\mathsf{C} = \mathsf{P}^{-1} \cdot \mathsf{S}. \tag{B.19}$$

When there are no sources $\mathsf{S} = 0$, and in order for a nontrivial solution of Equation B.17 to exist,

$$\det \mathsf{P} = 0. \tag{B.20}$$

Equation B.20 defines an eigenvalue problem that may be solved for the critical volume ($R_0$ and $\tilde{a}$) or the critical composition ($\alpha_{\pm}$, $\beta_{\pm}$). In this case, $C_1$, $D_1$, and $H_1$ may be evaluated as multiples of $A_1$, which latter may be chosen to satisfy a normalization condition or left arbitrary. More complicated source distributions do not affect the form of the result but merely change the form of the functions $x_1$ and $x_2$.

The computational effort required to solve the two-energy mode, two-region problem is the same as that required for the two-group, two-region problem. The extension of the solutions of Equation B.1 to two-dimensional problems should follow the same procedure that has been developed for the two-group equations.[39] Extension of the analytical solution to one-dimensional problems having more than two modes or two regions is quite straightforward, but the results are too complex to be useful. In a $N$-mode, $M$-region problem, $M$-1 matrix equations of rank $2N$ must be solved simultaneously. These arise from the $2N$ interface boundary conditions at the $M$-1 interfaces. For such problems, the finite difference method developed in the next appendix is more feasible.

# Appendix C. Energy Modal Finite Difference Equations

For problems in which the neutron distribution depends upon only one spatial variable ($r$), the energy modal equation, Equation 3.30, may be written

$$-\frac{1}{r^\beta}\left[\frac{d}{dr}\left(r^\beta \mathsf{D}(r)\cdot\frac{d}{dr}\rho(r)\right)\right] + \Sigma(r)\cdot\rho(r) = \mathsf{f}(r), \tag{C.1}$$

where

$$\beta = \begin{matrix} 0, \text{ slab,} \\ 1, \text{ cylinder,} \\ 2, \text{ sphere.} \end{matrix}$$

The interval over which $r$ varies may be broken up into a finite number $(K + 1)$ of discrete intervals. These are chosen such that any change in material properties occurs at a mesh point.

$$\rightarrow|\ h_n\ |\leftarrow$$

$$0 \quad 1 \quad 2 \quad 3 \quad n-1 \quad n \quad n+1 \quad n+2 \quad K-1 \quad K \quad K+1$$

Denote the constant values of the diffusion coefficient, cross section, and source matrices in the interval $n \rightarrow n + 1$ by the subscript $n$. Integrate from $n - \frac{1}{2}$ to $n + \frac{1}{2}$, use a central difference approximation for the derivative term, and approximate the integrals involving $\Sigma \cdot \rho$

and f by a one-term Taylor's series. Under these operations, Equation C.1 becomes

$$\mathsf{d}_n \cdot \rho_n = \omega_n + \frac{r_n{}^\beta h_n}{2}\mathsf{f}_n + \frac{r_n{}^\beta h_{n-1}}{2}\mathsf{f}_{n-1}, \qquad n = 1,\cdots, K, \quad (C.2)$$

where

$$\omega_n \equiv \mathsf{a}_n \cdot \rho_{n+1} - (\mathsf{a}_n + \mathsf{a}_{n-1}) \cdot \rho_n + \mathsf{a}_{n-1} \cdot \rho_{n-1}, \quad (C.3)$$

and

$$\mathsf{a}_n \equiv \frac{r^\beta_{n+1/2}}{h_n}\mathsf{D}_n, \qquad \mathsf{d}_n \equiv \frac{h_{n-1} r_n{}^\beta}{2}\Sigma_{n-1} + \frac{h_n r_n{}^\beta}{2}\Sigma_n. \quad (C.4)$$

Equations C.2 and C.3 apply at all internal mesh points $1 \leqslant n \leqslant K$. Equation C.2 is of the form of a balance equation on node $n$. These equations may be solved by standard methods.[9]

In order to solve these equations, which apply only at interior mesh points, $\rho_0$ and $\rho_{K+1}$ must be specified by boundary conditions. The zero flux boundary condition is $\rho_{K+1} = 0$ or $\rho_0 = 0$, while a zero current boundary condition at $n = 0$ takes the form

$$\frac{d}{dr}\rho_0 \Rightarrow \frac{\rho_1 - \rho_0}{h_0} = 0, \qquad \text{or } \rho_0 = \rho_1.$$

Equations C.2 may be written as a matrix equation involving matrices of rank $K \times N$ where $N$ is the number of energy modes. The conventional multigroup finite difference representation with $M$ groups involves matrices of rank $K \times M$.

# Appendix D. Evaporation Model of Inelastic Spectra

The statistical theory of the nucleus[24] is based upon five assumptions concerning the interaction of a nucleus and incident nucleon:

1. The incident nucleon and original nucleus form a compound nucleus that lasts for a time which may be considered as long relative to the natural period of the nucleon within the nuclear dimension.
2. The excitation energy of the compound nucleus is sufficiently great that a continuum of levels is formed.
3. The mode of decay of the compound nucleus is independent of its mode of formation, depending only upon the good quantum numbers of the system.
4. For each possible mode of decay, the states of the residual nucleus-decay product system form a continuum of levels.
5. Compound elastic scattering is negligible.

These assumptions and the predictions of the statistical theory are discussed in Reference 24, where it is shown that the spectrum of inelastically scattered neutrons may be represented by the evaporation model; that is,

$$P(E' \rightarrow E) = \frac{Ee^{-E/T(E')}}{T^2(E')}, \tag{D.1}$$

where $E$ and $E'$ are the energies of the emitted and incident neutrons, respectively. Expressions for $T$ have been derived from various theoretical models of the nucleus. Two of the simpler and more widely used relations are

$$T = \left[\left(\frac{\mu A}{U}\right)^{1/2} - \frac{2}{U}\right]^{-1}, \quad \text{Bethe,} \tag{D.2}$$

$$T = \left(\frac{U}{\nu A}\right)^{1/2}, \quad \text{Weisskopf,} \tag{D.3}$$

where $U$ is the *maximum* excitation energy of the *residual* nucleus, $A$ is the atomic mass number, and $\mu$ and $\nu$ are parameters related to the nuclear level spacing.

Empirical values of $T$ have been determined from experimental data. A least-squares fit of level spacing data for many nuclei with the formulas from which Equations D.2 and D.3 were derived led to "best" values of the level spacing parameters of $\nu = 0.0888$ Mev$^{-1}$ and $\mu = 0.1023$ Mev$^{-1}$.[64] Both $\nu = 0.160$ Mev$^{-1}$ and $\mu = 0.188$ Mev$^{-1}$ minimize a weighted deviation between Equations D.2 and D.3 and the experimental data for many nuclei;[62] the experimental uncertainty was the weighting factor. The authors of the YOM cross-section set,[68] which is widely used for fast-reactor calculations, chose Equation D.3 with $\nu = 0.0961$ Mev$^{-1}$ to evaluate inelastic transfer matrix elements. An experimental determination of the parameter $T$ for $U^{238}$ [62] may be made by fitting Equation D.3, which yields $\nu = 0.113$ Mev$^{-1}$.

This plethora of "best" empirical values for the level spacing parameter indicates that some ambiguity must accompany the specification of the parameter $T$ in Equation D.1. The approach that is adopted can be summarized as follows:

a. Use a single value $T_1$ for all uranium and plutonium isotopes and a different value $T_2$ for all structural, coolant, and other dilutant nuclei.
b. Use one of the previous formulas to determine $T_1$ and $T_2$. Replace $U$ by the average incident energy for neutrons undergoing inelastic collisions. This average value must be estimated, and is, in general, different for fuel nuclei and dilutant nuclei.

The highly enriched assemblies of Section 4.3 are evaluated for $T = 0.29$, 0.33, and 0.39 Mev. These values seem to span the range of possibilities and may be motivated by several arguments, three of which follow:

a. Use Equation D.3 and $\nu = 0.0888$ Mev$^{-1}$. For average incident neutron energies of 1.8, 2.3, and 3.25 Mev, the corresponding values of $T$ are 0.29, 0.33, and 0.39 Mev.

b. Use Equation D.2 and assume an average incident neutron energy of 2.5 Mev. For $\mu = 0.1023$ and $0.188 \text{ Mev}^{-1}$, the parameter $T$ is 0.29 and 0.39 Mev, respectively.
c. Use Equation D.3 with $v = 0.160 \text{ Mev}^{-1}$, and assume the average incident neutron energy is 2.5 Mev. This yields $T = 0.33$ Mev.

The dilute assemblies of Section 4.4 are evaluated for $T_1 = 0.16$ and 0.26 Mev, and for $T_2 = 1.5$ Mev. The values of $T_1$ were obtained by using Equation D.3 with the experimentally determined value of the level spacing parameter for $U^{238}$ of $v = 0.113 \text{ Mev}^{-1}$. Average incident neutron energies of 0.69 and 1.82 Mev, which seem to bracket the range of possibilities in this type assembly, then lead to $T = 0.16$ and 0.26 Mev, respectively. The value $T_2 = 1.5$ Mev was obtained by estimating the best fit of Equation D.1 to measured excitation functions for Na, Al, Fe, Ni, and Cr,[25] which are the principle inelastic scattering dilutants found in dilute fast-reactor assemblies.

# Appendix E. Spectrum and Importance Function

In Chapter 4, analytical expressions for the spatially independent spectrum and importance function were developed by solving the appropriate equations by the method of successive approximations, and then making certain approximations to the iterated kernels. The purposes of this appendix are to derive the general expressions for the spectrum and importance function from physical arguments and to present the details of the approximations that are used to reduce these general solutions to simple analytical expressions.

## E.1 Spectrum — Physical Derivation

The stationary collision density at energy $E$ may be constructed from the principle of neutron balance, which states that the rate at which neutrons are removed from an incremental energy element $dE$ is equal to the rate at which neutrons are being introduced into $dE$. Neutrons are removed by scattering, absorption, and leakage; the total removal rate is defined as the collision density, $F(E)$, times $dE$. Neutrons are introduced into $dE$ by the following processes in a source-free region:

1. Virgin fission neutrons born in $dE$.
2. Neutrons born of fission at $E' > E$ that undergo $n$ successive elastic scattering events, the $n$th event resulting in neutrons that have energy within $dE$ ($1 \geqslant n \geqslant \infty$).

3. Neutrons born of fission at $E'$ that undergo all possible sequences of elastic *and inelastic* events that could result in a neutron having energy within $dE$.

The neutron balance principle may thus be written

$$F(E) = [1] + [2] + [3]. \tag{E.1}$$

The quantity that belongs in the first brackets is, obviously, the fission spectrum. In evaluating the individual terms in the second bracket, it is necessary to consider all nonelastic events to remove a neutron from consideration, and pose the question: What is the distribution of fission neutrons after $n$ events?

$$\bar{K}_1(E) \equiv \int_E^{E/\alpha} \frac{r_{et}(E')\chi(E')}{E'(1-\alpha)}\, dE' \tag{E.2}$$

is the distribution of fission neutrons after one event, with the assumption that all nonelastic events remove a neutron from consideration. It follows that

$$\bar{K}_n(E) \equiv \int_E^{E/\alpha} \frac{r_{et}(E')\bar{K}_{n-1}(E')}{E'(1-\alpha)}\, dE', \tag{E.3}$$

is the distribution of fission neutrons after $n$ events, and $\bar{K}_n(E)dE$ is the number of neutrons within $dE$ about $E$ which were born of fission and then underwent $n$ successive elastic collisions. Thus, the term in the second brackets is

$$[2] = \sum_{n=1}^{\infty} \bar{K}_n(E). \tag{E.4}$$

Specification of the quantity in the third brackets is more complex, because it consists of a multiple infinity of terms that account for all possible sequences involving elastic and inelastic events. Consider the term that describes those neutrons which are born in the fission spectrum, undergo $n$ successive elastic events, then have an inelastic collision with a nucleus of species $m$, then undergo $n'$ successive elastic events followed by an inelastic collision with a nucleus of species $m'$, then have $n''$ successive elastic collisions. The $n''$th collision leaves the neutron with energy within $dE$ about $E$.

It is assumed that the inelastic spectrum is independent of the incident neutron energy, which is consistent with the model used in Chapter 4 to describe the inelastic spectrum. This has the effect of making the inelastic spectrum independent of the neutron's history.

The number of fission neutrons that suffer $n$ successive elastic events and obtain energy within $dE'$ about $E'$ on the $n$th one is given by Equation E.3. Thus, the distribution of fission neutrons that suffer $n$ successive elastic collisions and then have an inelastic collision with a nucleus of species $m$ is

$$P^m(E'') \int_0^\infty dE' \bar{K}_n(E') r_{in}^m(E'). \tag{E.5}$$

The same reasoning that led to Equation E.3 argues that the distribution of these neutrons which then have $n'$ successive elastic events is

$$\bar{J}_{n'}{}^m(E'') \cdot \int_0^\infty dE' \bar{K}_n(E') r_{in}^m(E'), \tag{E.6}$$

where

$$\bar{J}_1{}^m(E) \equiv \int_E^{E/\alpha} \frac{r_{et}(E') P^m(E')}{E'(1-\alpha)} dE', \tag{E.7}$$

and

$$\bar{J}_n{}^m(E) \equiv \int_E^{E/\alpha} \frac{r_{et}(E') \bar{J}_{n-1}^m(E')}{E'(1-\alpha)} dE', \qquad n \geqslant 2. \tag{E.8}$$

Consequently, the distribution of fission neutrons that suffer $n$ successive elastic events, then have an inelastic collision with a nucleus of species $m$, then suffer $n'$ successive elastic events, and then have an inelastic collision with a nucleus of species $m'$ is

$$P^{m'}(E''') \cdot \int_0^\infty dE'' \bar{J}_{n'}^m(E'') r_{in}^{m'}(E'') \cdot \int_0^\infty dE' \bar{K}_n(E') r_{in}^m(E').$$

Finally, the distribution of these neutrons that subsequently have $n''$ successive elastic events is obtained by the same argument that led to Equation E.6, and is

$$\bar{J}_{n''}^{m'}(E) \cdot \int_0^\infty dE'' \bar{J}_{n'}^m(E'') r_{in}^{m'}(E'') \cdot \int_0^\infty dE' \bar{K}_n(E') r_{in}^m(E'). \tag{E.9}$$

The integrals in Expression E.9 are constants which may be evaluated by using Equations E.3 and E.8. These constants are defined as

$$r_n{}^m \equiv \int_0^\infty dE r_{in}^m(E) \bar{K}_{n-1}(E), \tag{E.10}$$

$$l_n{}^{m'm} \equiv \int_0^\infty dE r_{in}^{m'}(E) \bar{J}_{n-1}^m(E). \tag{E.11}$$

$r_n{}^m$ has the physical significance of being the probability that a neutron born of fission has $n - 1$ successive elastic collisions then has an inelastic collision with a nucleus of species $m$, while $l_n{}^{m'm}$ is the probability that a neutron that is inelastically scattered from a nucleus of species $m$ has $n - 1$ successive elastic collisions followed by an inelastic collision with a nucleus of species $m'$. With these definitions, Expression E.9 becomes

$$\bar{J}_{n''}^{m'}(E) \cdot r_n{}^m \cdot l_{n'}^{m'm}. \tag{E.12}$$

Construction of other sequences follows this same procedure. A sequence involving three inelastic events, for instance, would be similar to Expression E.12 except that it would contain an additional $l^{m''m}$ term. The quantity that belongs in the third brackets may thus be written by considering all possible sequences of this nature

$$[3] = \sum_{m=1}^{M} \beta^m \left[ P^m(E) + \sum_{n=1}^{\infty} \bar{J}_n{}^m(E) \right], \tag{E.13}$$

$$\beta^m \equiv r^m + \sum_{i=1}^{M} r^i l^{mi} + \sum_{j=1}^{M} \sum_{i=1}^{M} r^i l^{ji} l^{mj}$$
$$+ \sum_{k=1}^{M} \sum_{j=1}^{M} \sum_{i=1}^{M} r^i l^{ji} l^{kj} l^{mk} + \cdots. \tag{E.14}$$

The following definitions have been made:

$$r^m \equiv \sum_{n=1}^{\infty} r_n{}^m, \qquad l^{mi} \equiv \sum_{n=1}^{\infty} l_n{}^{mi}. \tag{E.15}$$

The first term in $\beta^m$ accounts for all sequences involving a single inelastic event, the second term accounts for all sequences involving two inelastic events, etc. Here $M$ is the number of different inelastic scattering species present.

The collision density may now be constructed according to Equation E.1.

$$F(E) = [\chi(E)] + \left[ \sum_{n=1}^{\infty} \bar{K}_n(E) \right] + \sum_{m=1}^{M} \beta^m \left[ P^m(E) + \sum_{n=1}^{\infty} \bar{J}_n{}^m(E) \right]. \tag{E.16}$$

Equation E.16 is identical in every respect to Equation 4.6, which was found by solving the neutron balance equation by the method of successive approximations.

## E.2 Importance Function—Physical Derivation

The physical derivation of an expression for the importance function in an infinite critical medium is based upon its interpretation as an iterated fission probability.[30] This viewpoint visualizes the importance function as the asymptotic power level that results when a neutron is introduced at a given point, with a given energy, into a critical reactor which is initially at zero power. For an *infinite critical* medium, this asymptotic power level is proportional to the number of fission neutrons produced directly by the original neutron. This number of fission neutrons must, of course, be a statistical average.

Hence, the importance function for an infinite critical medium can be constructed by evaluating the probability that $\nu$ neutrons are produced by fission following a given sequence of elastic and inelastic events, and then summing over all such sequences. It is convenient to consider three types of sequences by which a neutron at energy $E$ can produce $\nu$ neutrons by fission:

1. Fission at $E$ without an intervening scattering event.
2. Fission at $E' < E$ following $n$ successive elastic collisions ($1 \leqslant n \leqslant \infty$).
3. Fission at $E'$ following a sequence of elastic *and inelastic* events.

The importance function may be written

$$\phi^*(E) = [1] + [2] + [3]. \tag{E.17}$$

The quantity in the first brackets is obviously just $\nu$ times the fission probability at $E$,

$$[1] = \nu r_f(E) \equiv \bar{k}_1(E).$$

The probability that a neutron at $E$ has one elastic collision and then produces $\nu$ fission neutrons is

$$\bar{k}_2(E) = \frac{r_{et}(E)}{E(1-\alpha)} \int_{\alpha E}^{E} \nu r_f(E')dE', \tag{E.18}$$

and the probability that a neutron at $E$ has $n-1$ successive elastic collisions and then produces $\nu$ fission neutrons is

$$\bar{k}_n(E) = \frac{r_{et}(E)}{E(1-\alpha)} \int_{\alpha E}^{E} \bar{k}_{n-1}(E')dE', \qquad n \geqslant 2. \tag{E.19}$$

The quantity in the second brackets of Equation E.17 is thus

$$[2] = \sum_{n=2}^{\infty} \bar{k}_n(E). \tag{E.20}$$

From the same type of reasoning which led to Equation E.19, it follows that the probability that a neutron at $E$ has $n-1$ successive elastic events followed by an inelastic collision with a nucleus of species $m$ is

$$\bar{h}_n{}^m(E) = \frac{r_{et}(E)}{E(1-\alpha)} \int_{\alpha E}^{E} \bar{h}^m_{n-1}(E')\,dE', \qquad n \geqslant 2, \tag{E.21}$$

where

$$\bar{h}_1{}^m(E) = r^m_{in}(E). \tag{E.22}$$

Two other probabilities are necessary in order to construct a sequence for the third brackets of Equation E.17. The probability that a neutron, which has scattered inelastically from a nucleus of species $m$, has $n-1$ subsequent elastic collisions and then produces $\nu$ neutrons by fission is evaluated by averaging $\bar{k}_n$ over the inelastic spectrum of species $m$.

$$f_n{}^m \equiv \int_0^\infty dEP^m(E)\bar{k}_n(E), \qquad n \geqslant 1. \tag{E.23}$$

Similarly, the probability that a neutron, which has scattered inelastically from a nucleus of species $m$, has $n-1$ subsequent elastic collisions followed by an inelastic collision with a nucleus of species $m'$ is evaluated by averaging $\bar{h}_n{}^{m'}$ over the inelastic spectrum of species $m$.

$$g_n{}^{mm'} \equiv \int_0^\infty dEP^m(E)\bar{h}_n{}^{m'}(E), \qquad n \geqslant 1. \tag{E.24}$$

A typical sequence from those implied by the third brackets of Equation E.17 can now be constructed, The probability that a neutron at $E$ has $n-1$ elastic events, then has an inelastic collision with a nucleus of species $m$, then has $n'-1$ successive elastic events, then has an inelastic collision with a nucleus of species $m'$, then has $n''-1$ successive elastic events, and then produces $\nu$ neutrons by fission is

$$\bar{h}_n{}^m(E)\cdot g^{mm'}_{n'}\cdot f^{m'}_{n''}.$$

The quantity in the third square brackets is constructed by considering all such sequences.

$$[3] = \sum_{m=1}^{M}\left[\sum_{n=1}^{\infty} f_n{}^m\right]\cdot\left[\sum_{n=1}^{\infty}\bar{h}_n{}^m(E) + \sum_{m'=1}^{M}\left(\sum_{n=1}^{\infty}\bar{h}_n{}^{m'}(E)\right)\left(\sum_{n=1}^{\infty} g_n{}^{m'm}\right)\right.$$
$$\left. + \sum_{m''=1}^{M}\sum_{m'=1}^{M}\left(\sum_{n=1}^{\infty}\bar{h}_n{}^{m'}(E)\right)\left(\sum_{n=1}^{\infty} g_n{}^{m'm''}\right)\left(\sum_{n=1}^{\infty} g_n{}^{m''m}\right) + \cdots\right]. \tag{E.25}$$

In Equation E.25, the first square brackets contains the probability that a neutron which has scattered inelastically from a nucleus of species

$m$ will produce $\nu$ neutrons by fission without an intervening inelastic event. The second square brackets contains the cumulative probability that a neutron at $E$ will have an inelastic collision with a nucleus of species $m$. Another way of interpreting Equation E.25 is to remove the brackets and multiply the $f_n{}^m$ series into all the series in the second brackets. Then the first term of this new expression,

$$\sum_{m=1}^{M} \left(\sum_{n=1}^{\infty} f_n{}^m\right)\left(\sum_{n=1}^{\infty} \bar{h}_n{}^m(E)\right),$$

is the probability that a neutron at $E$ produces $\nu$ fission neutrons with a single intervening inelastic collision. The second term,

$$\sum_{m=1}^{M} \left(\sum_{n=1}^{\infty} f_n{}^m\right)\left(\sum_{m'=1}^{M} \left\{\sum_{n=1}^{\infty} \bar{h}_n{}^{m'}(E)\right\}\left\{\sum_{n=1}^{\infty} g_n{}^{m'm}\right\}\right),$$

is the probabliity that a neutron at $E$ produces $\nu$ fission neutrons with two intervening inelastic collisions, etc.

When these results are substituted into Equation E.17, the resulting equation is identical in every aspect to Equation 4.22, which was derived by solving the conservation of importance equation by the method of successive approximations.

### E.3 Effective Scattering and Fission Probabilities

In Chapter 4, it was necessary to specify effective scattering and fission probabilities in order that the expressions for the spectrum and importance function could be put into a reasonably simple form by summing certain series. The purpose of this section is to arrive at prescriptions for these effective probabilities that may be readily evaluated. Physical arguments, based on the derivations of Sections E.1 and E.2, are employed to motivate qualitative definitions. Prescriptions are then obtained by using continuous slowing down theory to express these qualitative definitions in mathematical form.

The basic assumptions of continuous slowing down theory[23,59] are (1) that $E'F(E')$ is a slowly varying function in the interval $E/\alpha \geqslant E' \geqslant E$, and (2) the elastic scattering probability is constant and almost equal to unity in this interval. Elastic scattering resonances, inelastic scattering, and the presence of virgin fission neutrons make these assumptions of dubious validity in fast-reactor assemblies. However, it appears that the functional flux shape predicted by continuous slowing down theory is adequate to characterize the spectrum in dilute fast assemblies.[17] Also, in the range $0.9 > E > 0.01$ Mev, where the major portion of the spectrum is located in fast assemblies, the elastic scattering probability is on the order of 0.85-0.95. The heavy and intermediate mass nuclei, which predominate in many fast assemblies, lend credence to the use of continuous slowing down theory. These factors, together with the simplicity of the formalism, encouraged the use of continuous slowing down theory.

The results from continuous slowing down theory which shall be employed are:

1. The number of collisions that a neutron suffers in elastically moderating over a differential energy increment $dE$ is

$$d\left(\frac{\ln E}{\xi}\right),$$

where $\xi$ is the mean logarithmic energy decrement,

$$\xi = 1 + \frac{\alpha \ln \alpha}{1 - \alpha}.$$

2. The probability that a neutron at $E'$ moderates to $E$ without a nonelastic collision is

$$p(E'/E) = \exp\left[-\int_E^{E'} \frac{1 - r_{et}(E'')}{\xi(E'')} \cdot \frac{dE''}{E''}\right]. \tag{E.26}$$

This is a nonelastic escape probability that reduces to the familiar resonance escape probability when inelastic scattering is absent.

The formal definition of the effective elastic scattering probability follows from Equation 4.11, which may be written

$$(\bar{r}_{et}(E))^n \equiv \frac{\overline{K}_n(E)}{K_n(E)}$$
$$= \frac{\displaystyle\int_E^{E/\alpha} \frac{r_{et}(E')\,dE'}{E'(1-\alpha)} \int_{E'}^{E'/\alpha} \frac{r_{et}(E'')\,dE''}{E''(1-\alpha)} \cdots \int_{E^{(n-1)}}^{E^{(n-1)}/\alpha} \frac{r_{et}(E^{(n)})\chi(E^{(n)})\,dE^{(n)}}{E^{(n)}(1-\alpha)}}{\displaystyle\int_E^{E/\alpha} \frac{dE'}{E'(1-\alpha)} \int_{E'}^{E'/\alpha} \frac{dE''}{E''(1-\alpha)} \cdots \int_{E^{(n-1)}}^{E^{(n-1)}/\alpha} \frac{\chi(E^{(n)})\,dE^{(n)}}{E^{(n)}(1-\alpha)}}. \tag{E.27}$$

It was shown in Section E.1 that $\overline{K}_n$ is the distribution of fission neutrons after $n$ events, assuming that nonelastic events remove a neutron from consideration. Obviously $K_n$ is the distribution of fission neutrons after $n$ events if the elastic scattering probability is unity. Thus, $\bar{r}_{et}$ in Equation E.27 may be interpreted as the average elastic scattering probability experienced by a neutron in moderating from the fission spectrum to $E$ in $n$ elastic collisions. Because it is required that Equation E.27 be satisfied by a single $\bar{r}_{et}(E)$ for all values of $n$, $\bar{r}_{et}$ must be interpreted as the average elastic scattering probability experienced by a neutron in moderating from the fission spectrum in $n$ elastic collisions, weighted by the relative number of neutrons at $E$ which have moderated from the fission spectrum in $n$ collisions, and averaged over all values of $n$. This is equivalent to the average value of the elastic scattering probability that has been experienced by neutrons at $E$ in elastically moderating from the fission spectrum. The average elastic scattering probability experienced by a neutron in elastically moderating from $E_f$ to $E$ is

$$\langle r_{et}(E_f/E)\rangle \equiv \frac{\int_E^{E_f} r_{et}(E')\, d\left(\frac{\ln E'}{\xi}\right)}{\int_E^{E_f} d\left(\frac{\ln E'}{\xi}\right)}. \tag{E.28}$$

This quantity is constructed by weighting the elastic scattering probability at each energy $E'$ by the number of collisions required to traverse an incremental energy interval $dE'$ about energy $E'$.

The relative number of neutrons at energy $E$ which have elastically moderated from energy $E_f$ in the fission spectrum is proportional to the relative number of neutrons originating at energy $E_f$, which is $\chi(E_f)$, and to the relative probability that a neutron at energy $E_f$ moderates to energy $E$ without a nonelastic collision, which is $p(E_f/E)$ of Equation E.26. Thus, the effective elastic scattering probability $\bar{r}_{et}$ may be written

$$\bar{r}_{et}(E) = \frac{\int_E^{\infty} dE_f \chi(E_f) p(E_f/E)\langle r_{et}(E_f/E)\rangle}{\int_E^{\infty} dE_f \chi(E_f) p(E_f/E)}. \tag{E.29}$$

A prescription for $\bar{r}_{et}^m$ ,which must satisfy Relation 4.12, is developed by analogous arguments. The only difference being that neutrons are moderating from the inelastic spectrum $P^m(E)$ rather than from the fission spectrum.

$$\bar{r}_{et}^m(E) = \frac{\int_E^{\infty} dE_f P^m(E_f) p(E_f/E)\langle r_{et}(E_f/E)\rangle}{\int_E^{\infty} dE_f P^m(E_f) p(E_f/E)}. \tag{E.30}$$

Iterated kernels in the expression for the importance function were also approximated by specifying effective probabilities. To investigate these probabilities, the first of Equations 4.26 is rewritten as

$$(\tilde{r}_{et}(E))^{n-1} \cdot \widetilde{\nu r}_f(E) = \frac{\bar{k}_n(E)}{k_n(E)}$$

$$= \frac{\dfrac{r_{et}(E)}{E(1-\alpha)} \displaystyle\int_{\alpha E}^{E} \frac{r_{et}(E')\, dE'}{E'(1-\alpha)} \cdots \int_{\alpha E^{(n-2)}}^{E^{(n-2)}} \nu r_f(E^{(n-1)})\, dE^{(n-1)}}{\dfrac{1}{E(1-\alpha)} \displaystyle\int_{\alpha E}^{E} \frac{dE'}{E'(1-\alpha)} \cdots \int_{\alpha E^{(n-2)}}^{E^{(n-2)}} dE^{(n-1)}}. \tag{E.31}$$

In Section E.2, $\bar{k}_n$ was shown to be the probability that a neutron at $E$ will have $n-1$ successive elastic collisions and then produce $\nu$ neutrons by fission. The denominator of Equation E.31 defines $k_n$ which is numerically equal to unity and which may be interpreted as the probability that a neutron at energy $E$ will have $n-1$ successive elastic collisions if the elastic scattering probability is unity. Thus, $\tilde{r}_{et}$ in Equation E.31 has the physical significance of being the average value of the elastic scattering probability that will be experienced by a neutron at energy $E$ in $n-1$ subsequent elastic events. Similarly, $\widetilde{\nu r}_f$ is the average probability of producing $\nu$ fission neutrons that will be experienced by a neutron at energy $E$ after $n-1$ subsequent elastic collisions.

Because it is required that Relation E.31 be valid for a single value of $\tilde{r}_{et}(E)$ and a single value of $\widetilde{\nu r}_f(E)$, for all values of $n$, it is necessary to define these quantities as weighted averages. Thus, $\tilde{r}_{et}(E)$ is interpreted as the average value of the elastic scattering probability that will be experienced by a neutron at energy $E$ in $n-1$ subsequent elastic collisions, weighted with the relative probability that a neutron at energy $E$ has $n-1$ and only $n-1$ subsequent elastic collisions, and averaged over all possible values of $n$. Again, $\widetilde{\nu r}_f$ is interpreted as the average probability of producing $\nu$ fission neutrons that will be experienced by a neutron at energy $E$ after $n-1$ elastic collisions, weighted with the probability that a neutron at energy $E$ has $n-1$ and only $n-1$ elastic collisions before suffering a nonelastic collision, and averaged over all possible values of $n$.

This is equivalent to interpreting $\tilde{r}_{et}(E)$ as the average value of the elastic scattering probability that will be experienced by a neutron at energy $E$ before it has a nonelastic collision, and interpreting $\widetilde{\nu r}_f(E)$ as the average probability of producing $\nu$ fission neutrons that will be experienced by a neutron at energy $E$ when it has a nonelastic collision.

The continuous slowing down theory expression for the collision-weighted average value of the elastic scattering probability which will be experienced by a neutron in elastically moderating from energy $E$ to energy $E_s$ is

$$\langle r_{et}(E/E_s)\rangle = \frac{\displaystyle\int_{Es}^{E} r_{et}(E')\, d\left(\frac{\ln E'}{\xi}\right)}{\displaystyle\int_{Es}^{E} d\left(\frac{\ln E'}{\xi}\right)}. \tag{E.32}$$

The probability that a neutron at energy $E$ will have its first nonelastic collision at energy $E_s$ is the product of the probability that a neutron will moderate from energy $E$ to energy $E_s$ without a nonelastic collision, $p(E/E_s)$ of Equation E.26, and the probability that a neutron at energy $E_s$ has a nonelastic collision, $1 - r_{et}(E_s)$. Thus,

$$\tilde{r}_{et}(E) = \frac{\displaystyle\int_0^E dE_s[1 - r_{et}(E_s)][p(E/E_s)]\langle r_{et}(E/E_s)\rangle}{\displaystyle\int_0^E dE_s[1 - r_{et}(E_s)][p(E/E_s)]}. \tag{E.33}$$

and

$$\widetilde{\nu r}_f(E) = \frac{\displaystyle\int_0^E dE_s[1 - r_{et}(E_s)][p(E/E_s)][\nu r_f(E_s)]}{\displaystyle\int_0^E dE_s[1 - r_{et}(E_s)][p(E/E_s)]}. \tag{E.34}$$

When the second of Equations 4.26 is written as

$$(\tilde{r}_{et}(E))^{n-1} \cdot \tilde{r}_{in}^m(E) = \frac{\bar{h}_n{}^m(E)}{h_n{}^m(E)}$$

$$= \frac{\dfrac{r_{et}(E)}{E(1-\alpha)} \displaystyle\int_{\alpha E}^{E} \frac{r_{et}(E')dE'}{E'(1-\alpha)} \cdots \int_{\alpha E^{(n-2)}}^{E^{(n-2)}} dE^{(n-1)} r_{in}^m(E^{(n-1)})}{\dfrac{1}{E(1-\alpha)} \displaystyle\int_{\alpha E}^{E} \frac{dE'}{E'(1-\alpha)} \cdots \int_{\alpha E^{(n-2)}}^{E^{(n-2)}} dE^{(n-1)}}$$

and compared with Equation E.31, it is apparent that the same argument made previously applies to $\tilde{r}_{et}$ again, and an argument analogous to that made for $\widetilde{\nu r}_f$ can be made for $\tilde{r}_{in}^m$. Thus, $\tilde{r}_{in}^m$ is interpreted as the average inelastic scattering probability, for species $m$, that will be experienced by a neutron at energy $E$ when it has a nonelastic collision, and is given by

$$\tilde{r}_{in}^m(E) = \frac{\displaystyle\int_0^E dE_s[1 - r_{et}(E_s)][p(E/E_s)][r_{in}^m(E_s)]}{\displaystyle\int_0^E dE_s[1 - r_{et}(E_s)][p(E/E_s)]}. \tag{E.35}$$

It should be noted that the prescriptions given by Equations E.29, E.30, E.33, E.34, and E.35, and the arguments leading to them are quite general and are independent of any particular model of the neutron moderation process. Continuous slowing down theory is introduced into these prescriptions by the specification of $\langle r_{et} \rangle$ and $p(E/E')$ according to Equations E.26, E.28, and E.32.

## E.4 Evaluation of Iterated Kernels and Summation of Series

It was stated in Chapter 4 that the modified iterated kernels $K_n$ and $J_n^m$ could be evaluated analytically, and that this evaluation led to summable series involving these kernels. The details of the evaluation of $K_n$ and $J_n^m$ are presented in this section, and the sums involving these kernels in the expression for the collision density are performed.

The $K_n$ are defined by Equations 4.13,

$$K_0(E) = \chi(E) = \frac{2}{\sqrt{\pi}} \frac{E^{1/2}}{\theta^{3/2}} e^{-E/\theta},$$
$$K_n(E) = \int_E^{E/\alpha} \frac{K_{n-1}(E')dE'}{E'(1-\alpha)}, \qquad n \geqslant 1. \tag{E.36}$$

The first kernel is

$$K_1(E) = \frac{2}{\sqrt{\pi\theta}} \int_{E/\theta}^{E/\alpha\theta} \frac{d\left(\frac{E'}{\theta}\right)\left(\frac{E'}{\theta}\right)^{-1/2} e^{-E'/\theta}}{1-\alpha}$$
$$= \frac{2}{\sqrt{\pi\theta}} \left[\frac{\gamma\left(\frac{1}{2}, \frac{E}{\alpha\theta}\right) - \gamma\left(\frac{1}{2}, \frac{E}{\theta}\right)}{1-\alpha}\right],$$

where $\gamma$ is the function complementary to the incomplete gamma function.[1] It possesses a series expression and a recursion relation that are useful,

$$\gamma(a, z) \equiv \int_0^z e^{-t} t^{a-1}\, dt = z^a e^{-z} \Gamma(a) \sum_{n=0}^{\infty} \frac{z^n}{\Gamma(a+n+1)}, \tag{E.37}$$

and

$$\gamma(a+1, z) = a\gamma(a, z) - z^a e^{-z}, \tag{E.38}$$

where $\Gamma$ is the gamma function.[2]

By using the series representation of $\gamma$, the expression for $K_1$ may be written

$$K_1(E) = \frac{2}{\theta}\left(\frac{E}{\alpha\theta}\right)^{1/2} e^{-E/\theta} \sum_{n=0}^{\infty} \left(\frac{E}{\alpha\theta}\right)^n \left(\frac{e^{-\frac{(1-\alpha)}{\alpha}\frac{E}{\theta}} - \alpha^{n+1/2}}{(1-\alpha)\Gamma(n+\frac{3}{2})}\right).$$

For media composed of intermediate and high atomic mass nuclei, $\alpha$ is near unity, and a power series expansion in $(1 - \alpha)$ should converge rapidly. In dilute fast-reactor assemblies, $(1 - \alpha)$ is on the order of 0.1, and $K_1$ may be expanded in powers of $(1 - \alpha)$. By noting that

$$\begin{aligned}\frac{e^{-\frac{(1-\alpha)}{\alpha}\frac{E}{\theta}} - \alpha^{n+1/2}}{1-\alpha} &= \left[\left(n + \frac{1}{2}\right) - \left(\frac{E}{\alpha\theta}\right)\right] - \frac{(1-\alpha)}{2} \\ &\times \left[\left(n + \frac{1}{2}\right)\left(n - \frac{1}{2}\right) - \left(\frac{E}{\alpha\theta}\right)^2\right] + O((1-\alpha)^2),\end{aligned} \tag{E.39}$$

$$\left(\frac{1}{\alpha}\right)^m = 1 + m(1-\alpha) + \frac{m(m+1)}{2}(1-\alpha)^2 + \cdots, \tag{E.40}$$

and by making use of the identities

$$\sum_{n=0}^{\infty} \frac{\left(n + \frac{1}{2}\right)\left(\frac{E}{\alpha\theta}\right)^n - \left(\frac{E}{\alpha\theta}\right)^{n+1}}{\Gamma\left(n + \frac{3}{2}\right)} \equiv \frac{1}{\sqrt{\pi}}, \tag{E.41}$$

$$\sum_{n=0}^{\infty} \frac{\left(\frac{E}{\alpha\theta}\right)^{n+2} - \left(n + \frac{1}{2}\right)\left(n - \frac{1}{2}\right)\left(\frac{E}{\alpha\theta}\right)^n}{\Gamma\left(n + \frac{3}{2}\right)} \equiv \frac{1}{\sqrt{\pi}}\left[\frac{1}{2} - \frac{E}{\alpha\theta}\right], \tag{E.42}$$

the foregoing expression for $K_1$ becomes

$$K_1(E) = \chi(E)\left[1 + \frac{(1-\alpha)}{2}\left\{\frac{3}{2} - \frac{E}{\theta}\right\} + O((1-\alpha)^2)\right]. \tag{E.43}$$

The second iterated kernel is

$$\begin{aligned}K_2(E) &= \int_E^{E/\alpha} \frac{K_1(E')\,dE'}{E'(1-\alpha)} \\ &= \left[1 + \frac{3}{2}\cdot\frac{(1-\alpha)}{2}\right]K_1(E) - \frac{(1-\alpha)}{2} \\ &\quad \times \frac{2}{\sqrt{\pi}\theta}\int_{E/\theta}^{E/\alpha\theta} \frac{d\left(\frac{E'}{\theta}\right)\left(\frac{E'}{\theta}\right)^{1/2} e^{-E'/\theta}}{1-\alpha}.\end{aligned}$$

Equations E.37 and E.38 may be used to evaluate the integral,

$$\frac{2}{\sqrt{\pi\theta}} \int_{E/\theta}^{E/\alpha\theta} \frac{d\left(\frac{E'}{\theta}\right)\left(\frac{E'}{\theta}\right)^{1/2} e^{-E'/\theta}}{1-\alpha} = \frac{2}{\sqrt{\pi\theta}} \left[ \frac{\gamma\left(\frac{3}{2}, \frac{E}{\alpha\theta}\right) - \gamma\left(\frac{3}{2}, \frac{E}{\theta}\right)}{1-\alpha} \right]$$

$$= \frac{K_1(E)}{2} + \frac{2}{\sqrt{\pi\theta}} \left(\frac{E}{\alpha\theta}\right)^{1/2} e^{-E/\theta} \left[ \frac{\alpha^{1/2} - e^{-\frac{(1-\alpha)}{\alpha}\frac{E}{\theta}}}{1-\alpha} \right].$$

Equations E.39 and E.40 may be used to expand the last term, and $K_2$ becomes

$$K_2(E) = \chi(E)\left[1 + \frac{(1-\alpha)}{2}\left\{2\left(\frac{3}{2} - \frac{E}{\theta}\right)\right\} + O((1-\alpha)^2)\right]. \quad \text{(E.44)}$$

When this process is repeated for higher iterated kernels, it becomes apparent that the generic form of the $n$th iterated kernel is

$$K_n(E) = \chi(E)\left[1 + \frac{(1-\alpha)}{2}\left\{n\left(\frac{3}{2} - \frac{E}{\theta}\right)\right\} + O((1-\alpha)^2)\right]. \quad \text{(E.45)}$$

The $J_n{}^m$ are defined by Equations 4.14,

$$\begin{aligned} J_0{}^m(E) &\equiv P^m(E) = \frac{Ee^{-E/T_m}}{T_m{}^2}, \\ J_n{}^m(E) &\equiv \int_E^{E/\alpha} \frac{J_{n-1}^m(E')\,dE'}{E'(1-\alpha)}, \qquad n \geqslant 1. \end{aligned} \quad \text{(E.46)}$$

Evaluation of these kernels is straightforward, involving only Equations E.39 and E.40 and well-known integrals. The generic form of the $n$th iterated kernel is

$$J_n{}^m(E) = P^m(E)\left[1 + \frac{(1-\alpha)}{2}\left\{n\left(2 - \frac{E}{T_m}\right)\right\} + O((1-\alpha)^2)\right]. \quad \text{(E.47)}$$

Equation 4.6 may be written in terms of these modified iterated kernels by making use of Equations 4.11 and 4.12.

$$F(E) = \left[\sum_{n=0}^{\infty} (\bar{r}_{et}(E))^n K_n(E)\right] + \sum_{m=1}^{M} \beta^m \left[\sum_{n=0}^{\infty} (\bar{r}_{et}^m(E))^n J_n{}^m(E)\right].$$

When Equations E.45 and E.47 are substituted into this expression and the following well known sums[32] are used

$$\sum_{n=0}^{\infty} x^n = \frac{1}{1-x}, \qquad \sum_{n=1}^{\infty} nx^n = \frac{x}{(1-x)^2},$$

the expression given in Chapter 4 is obtained.

$$F(E) = K(E) + \sum_{m=1}^{M} \beta^m J^m(E), \tag{E.48}$$

where

$$\begin{aligned} K(E) &\equiv \sum_{n=0}^{\infty} (\bar{r}_{et})^n K_n \\ &= \chi(E)\left[\frac{1}{1 - \bar{r}_{et}(E)} + \frac{(1-\alpha)}{2}\left\{\frac{\bar{r}_{et}(E)}{(1 - \bar{r}_{et}(E))^2}\left(\frac{3}{2} - \frac{E}{\theta}\right)\right\} + O((1-\alpha)^2)\right], \end{aligned} \tag{E.49}$$

$$\begin{aligned} J^m(E) &\equiv \sum_{n=0}^{\infty} (\bar{r}_{et}^m)^n J_n{}^m \\ &= P^m(E)\left[\frac{1}{1 - \bar{r}_{et}^m(E)} + \frac{(1-\alpha)}{2}\left\{\frac{\bar{r}_{et}^m(E)}{(1 - \bar{r}_{et}^m(E))^2}\left(2 - \frac{E}{T_m}\right)\right\} + O((1-\alpha)^2)\right]. \end{aligned} \tag{E.50}$$

# Appendix F. Average Elastic Moderating Properties for Mixtures

When a medium consists of a mixture of several different types of nuclei, it is necessary to define average elastic moderating properties. These are defined in the usual fashion,[60] and are presented here for ready reference.

If $\Sigma_{es}^i$, $\mu_i$, and $A^i$ are the average elastic scattering cross section, the average cosine of the scattering angle in the center of mass system, and the atomic mass number of nuclear species $i$, respectively, then the following definitions obtain for the mixture:

1. Average consistent $P_1$ elastic scattering cross section in the transport approximation:

$$\Sigma_{et} = \sum_i \Sigma_{et}^i = \sum_i \Sigma_{es}^i(1 - \mu_i).$$

2. Average $\alpha$:

$$\alpha = \frac{\sum_i \Sigma_{et}^i \alpha_i}{\sum_i \Sigma_{et}^i}, \qquad \alpha_i \equiv \left(\frac{A^i - 1}{A^i + 1}\right)^2.$$

3. Average cosine of the scattering angle in CM system:

$$1 - \langle\mu\rangle = \sum_i \Sigma_{es}^i(1 - \mu_i)\Big/\sum_i \Sigma_{es}^i.$$

# Appendix G. $P_1$ Transport Approximation

The $P_1$ transport approximation that was used in Section 3.2 is developed in this appendix. For the sake of clarity, infinite-slab geometry is considered and the neutron distribution is assumed symmetric about the transverse axis. In this case, Equation 3.1 becomes

$$\mu \frac{d}{dz} \phi(z, E, \mu) + \Sigma_t(z, E)\phi(z, E, \mu)$$

$$= \int_E^\infty dE' \int_{\Omega'} d\boldsymbol{\Omega}' \Sigma_{es}(z, E'\boldsymbol{\Omega}'/E\boldsymbol{\Omega})\phi(z, E', \mu')$$

$$+ \int_E^\infty dE' \int_{\Omega'} d\boldsymbol{\Omega}' \Sigma_{in}(z, E'\boldsymbol{\Omega}'/E\boldsymbol{\Omega})\phi(z, E', \mu')$$

$$+ \chi(E) \int_0^\infty dE' \int_{\Omega'} d\boldsymbol{\Omega}' \frac{\nu\Sigma_f(z, E')}{4\pi} \phi(z, E', \mu'). \tag{G.1}$$

Here, $z$ is the spatial distance along the axis, and $\mu$ is the cosine of the angle the direction vector $\boldsymbol{\Omega}$ makes with this axis.

If the medium is isotropic, the scattering probability depends only upon the angle between the initial and final neutron directions and the neutron energy, and may be expanded in terms of the cosine of this angle $\mu_0$

$$\Sigma_{es}(z, E'\boldsymbol{\Omega}'/E\boldsymbol{\Omega}) = \sum_{l'=0}^{\infty} \frac{2l' + 1}{4\pi} P_{l'}(\mu_0)\Sigma_{es}^{l'}(z, E'/E), \tag{G.2}$$

which implies, by virtue of the orthogonality of the Legendre polynomials ($P_l$), that

$$\Sigma_{es}^{l'}(z, E'/E) = \int_{\Omega_0} d\boldsymbol{\Omega}_0 P_{l'}(\mu_0)\Sigma_{es}(z, E'\boldsymbol{\Omega}'/E\boldsymbol{\Omega}). \qquad \text{(G.3)}$$

Equation G.2 is next substituted into Equation G.1, and the vector addition relation

$$P_l(\mu_0) = P_l(\mu)P_l(\mu') + 2\sum_{\beta=1}^{l} \frac{(l-\beta)!}{(l+\beta)!} P_l{}^{\beta}(\mu)P_l{}^{\beta}(\mu') \cos \beta(\eta - \eta')$$

is used, where $P_l^{\beta}$ is the associated Legendre polynomial and $\eta$ is the azimuthal angle associated with the neutron direction vector $\boldsymbol{\Omega}$. This substitution results in

$$\mu \frac{d}{dz} \phi(z, E, \mu) + \Sigma_t(z, E)\phi(z, E, \mu)$$

$$= \sum_{l'=0}^{\infty} P_{l'}(\mu) \frac{2l'+1}{2} \int_E^{\infty} dE' \Sigma_{es}^{l'}(z, E'/E) \int_{\mu'} d\mu' P_{l'}(\mu')\phi(z, E', \mu')$$

$$+ \int_E^{\infty} dE' \frac{\Sigma_{\text{in}}(z, E'/E)}{2} \int_{\mu'} d\mu' \phi(z, E', \mu')$$

$$+ \chi(E) \int_0^{\infty} dE' \frac{\nu\Sigma_f(z, E')}{2} \int_{\mu'} d\mu' \phi(z, E', \mu'). \qquad \text{(G.4)}$$

Inelastic scattering is assumed isotropic in writing Equation G.4. Inelastic anisotropy could easily be accommodated by an expansion similar to that of Equation G.2.

A modal expansion of the neutron flux, with respect to the angular variable

$$\phi(z, E, \mu) = \sum_{m=0}^{\infty} \frac{2m+1}{4\pi} P_m(\mu)\phi_m(z, E) \qquad \text{(G.5)}$$

may be made. When this expansion is substituted into Equation G.4, and the resulting expression is operated on in turn by $P_l(\mu)$, $l = 0, 1, 2, \cdots$, and integrated over $-1 \leqslant \mu \leqslant 1$, an infinite set of equations obtains, the generic form of which is

$$\begin{aligned}
&\frac{l+1}{2l+1}\frac{d}{dz}\phi_{l+1}(z,E)+\frac{l}{2l+1}\frac{d}{dz}\phi_{l-1}(z,E)+\Sigma_t(z,E)\phi_l(z,E)\\
&\quad=\int_E^\infty dE'\Sigma_{es}^l(z,E'/E)\phi_l(z,E')\\
&\quad\quad+\delta_{l,0}\left[\int_E^\infty dE'\Sigma_{\rm in}(z,E'/E)\phi_0(z,E')\right.\\
&\quad\quad\left.+\chi(E)\int_0^\infty dE'\nu\Sigma_f(z,E')\phi_0(z,E')\right],\qquad l=0,1,2,\cdots. \qquad \text{(G.6)}
\end{aligned}$$

The consistent $P_1$ equations are obtained from the first two of these $P_L$ equations by setting $\phi_l = 0$ for $l > 1$.

$$\begin{aligned}
&\frac{d}{dz}\phi_1(z,E)+\Sigma_t(z,E)\phi_0(z,E)\\
&\qquad=\int_E^\infty dE'\Sigma_{es}^0(z,E'/E)\phi_0(z,E')\\
&\qquad\quad+\int_E^\infty dE'\Sigma_{\rm in}(z,E'/E)\phi_0(z,E')\\
&\qquad\quad+\chi(E)\int_0^\infty dE'\nu\Sigma_f(z,E')\phi_0(z,E'),\\
&\frac{1}{3}\frac{d}{dz}\phi_0(z,E)+\Sigma_t(z,E)\phi_1(z,E)=\int_E^\infty dE'\Sigma_{es}^1(z,E'/E)\phi_1(z,E').
\end{aligned} \qquad \text{(G.7)}$$

These equations are unchanged by adding identical terms to each side. Consequently, they may also be written

$$\begin{aligned}
&\frac{d}{dz}\phi_1(z,E)+[\Sigma_t(z,E)-\Sigma_{es}^1(z,E)]\phi_0(z,E)\\
&\qquad=\int_E^\infty dE'\Sigma_{es}^0(z,E'/E)\phi_0(z,E')-\Sigma_{es}^1(z,E)\phi_0(z,E)\\
&\qquad\quad+\int_E^\infty dE'\Sigma_{\rm in}(z,E'/E)\phi_0(z,E')\\
&\qquad\quad+\chi(E)\int_0^\infty dE'\nu\Sigma_f(z,E')\phi_0(z,E'),\\
&\frac{1}{3}\frac{d}{dz}\phi_0(z,E)+[\Sigma_t(z,E)-\Sigma_{es}^1(z,E)]\phi_1(z,E)\\
&\qquad=\int_E^\infty dE'\Sigma_{es}^1(z,E'/E)\phi_1(z,E')-\Sigma_{es}^1(z,E)\phi_1(z,E).
\end{aligned} \qquad \text{(G.8)}$$

In Equations G.8,

$$\Sigma_{es}^{1}(z, E) = \int_{0}^{\infty} dE' \Sigma_{es}^{1}(z, E/E'). \tag{G.9}$$

The transport approximation arises from the assumption that the anisotropic scattering takes place without a change in neutron energy: i.e.,

$$\Sigma_{es}^{1}(z, E'/E) = \Sigma_{es}^{1}(z, E') \cdot \delta(E' - E). \tag{G.10}$$

Elastic scattering from heavy nuclei, for which the anisotropy is most pronounced, satisfies this approximation rather well.

When Equation G.10 is used in Equations G.8, the consistent $P_1$ equations with the transport approximation are obtained:

$$\begin{aligned} \frac{d}{dz} \phi_1(z, E) + \Sigma_{tr}(z, E)\phi_0(z, E) &= \int_{E}^{\infty} dE' \Sigma_{et}(z, E'/E)\phi_0(z, E') \\ &+ \int_{E}^{\infty} dE' \Sigma_{\text{in}}(z, E'/E)\phi_0(z, E') \\ &+ \chi(E) \int_{0}^{\infty} dE' \nu \Sigma_f(z, E')\phi_0(z, E'), \\ \frac{1}{3}\frac{d}{dz} \phi_0(z, E) + \Sigma_{tr}(z, E)\phi_1(z, E) &= 0, \end{aligned} \tag{G.11}$$

where

$$\begin{aligned} \Sigma_{tr}(z, E) &\equiv \Sigma_t(z, E) - \Sigma_{es}^{1}(z, E), \\ \Sigma_{et}(z, E'/E) &\equiv \Sigma_{es}^{0}(z, E'/E) - \Sigma_{es}^{1}(z, E'/E). \end{aligned} \tag{G.12}$$

Equations G.11 may be combined to yield a single second-order differential equation. In the general case, this equation is Equation 3.10.

# Glossary of Symbols

**English Symbols**

| | |
|---|---|
| $d_e$ | extrapolation length in units of transport mean free paths |
| $E$ | energy |
| $F(E)$ | neutron collision density |
| $f$ | source |
| $f_n^{\ m}$ | parameter defined by Equation 4.25 |
| $G$ | kernel for effect of source on physical process |
| $g_n^{\ m'm}$ | parameter defined by Equation 4.25 |
| $\bar{h}_n^{\ m}$ | iterated kernel defined by Equation 4.24 |
| $H$, $H^*$, $H^{-1}$ | linear operator, its adjoint, and its inverse |
| $\bar{J}_n^{\ m}$ | iterated kernel defined by Equation 4.8 |
| $J_n^{\ m}$ | modified iterated kernel defined by Equation 4.14 |
| $J^m$ | function defined by Equation 4.17 |
| $K$ | function defined by Equation 4.16 |
| | linear operator |
| $\bar{K}_n$ | iterated kernel defined by Equation 4.7 |
| $K_n$ | modified iterated kernel defined by Equation 4.13 |
| $k_{eff}$ | effective multiplication constant |
| $\bar{k}_n$ | iterated kernel defined by Equation 4.23 |
| $L$ | Lagrangian variational functional |
| $l^{mk}$ | parameter defined by Equation 4.10 |
| $\mathbf{n}$ | unit vector normal to surface |
| $P^m$ | inelastic spectrum of species $m$ |
| $p$ | nonelastic escape probability |

| | |
|---|---|
| $\mathbf{r}$ | spatial position vector |
| $r_f$ | fission probability |
| $r^m$ | parameter defined by Equation 4.10 |
| $r_{et}$ | elastic scattering probability |
| $r_{in}$ | inelastic scattering probability |
| $T_m$ | parameter in inelastic spectrum formula |
| $U$ | step function |
| | maximum excitation energy |
| $u$ | arbitrary function |
| $v$ | arbitrary function |
| $x_i$ | independent variable |

## Greek Symbols

| | |
|---|---|
| $\alpha$ | $(A - 1)^2/(A + 1)^2$, $A$ = atomic mass number |
| $\alpha_j$ | weighting operator |
| $\alpha_\pm$ | parameter defined by Equation B.13 |
| $\beta_\pm$ | parameter defined by Equation B.16 |
| $\beta^m$ | parameter defined by Equation 4.9 |
| $\Gamma$ | gamma function |
| $\gamma$ | function complementary to incomplete gamma function |
| $\delta_{ij}$ | Kronecker delta |
| $\delta$ | Dirac delta function |
| $\theta$ | parameter in fission spectrum formula |
| $\lambda$ | control parameter, eigenvalue |
| $\mu$ | cosine of angle $\boldsymbol{\Omega}$ makes with polar axis |
| | level spacing parameter |
| $\nu$ | number of neutrons produced per fission |
| | level spacing parameter |
| $\xi$ | mean logarithmic energy decrement |
| $\xi_P$ | iteration parameter defined by Equation 5.17 |
| $\rho_i$ | modal combining coefficient |
| $\Sigma$ | macroscopic cross section |
| $\sigma$ | microscopic cross section |
| $\bar{\sigma}$ | reaction rate corresponding to $\sigma : \int \sigma\phi \, dE$ |
| $\phi$ | solution |
| | neutron flux |
| $\phi^*$ | importance (adjoint) function |
| $\chi$ | fission spectrum |
| $\psi_i$ | mode |
| $\psi_i{}^*$ | expansion function for adjoint |
| $\boldsymbol{\Omega}$ | unit directional vector |

### Mathematical Symbols

| | |
|---|---|
| $\nabla$ | del operator |
| $\nabla^2$ | Laplacian operator |
| $\delta$ | variation operator |
| $\sum$ | summation operator |
| $\int$ | integration operator |
| $\langle\ \rangle$ | averaging operator |
| $\|\psi\rangle$ | vector |
| $\langle\alpha\|$ | dual vector |
| $\langle\alpha\|\phi\rangle$ | $= \int dx\alpha\phi$ |
| $\langle\alpha\|H\|\phi\rangle$ | $= \int dx\alpha H\phi$ |

### Subscripts and Superscripts

| | |
|---|---|
| *es* | elastic scattering |
| *et* | $P_1$ elastic scattering in transport approximation |
| *f* | fission |
| *in* | inelastic scattering |
| *L* | leakage |
| *ne* | nonelastic |
| *t* | total |
| *tr* | transport |
| 23 | $U^{233}$ |
| 24 | $U^{234}$ |
| 25 | $U^{235}$ |
| 26 | $U^{236}$ |
| 28 | $U^{238}$ |
| 37 | $Np^{237}$ |
| 40 | $Pu^{240}$ |
| 49 | $Pu^{239}$ |

### Other Notation

Boldface type is used to denote vectors, and sans serif type refers to a matrix.

# References


1. Abramowitz, M., and I. A. Stegun, Eds., *Handbook of Mathematical Functions*, Dover, New York (1965), p. 260.
2. Abramowitz, M., and I. A. Stegun, *ibid.*, p. 255.
3. Bates, J. C., *et al.*, "Two Spherical Fast Critical Experiments," ANL-6582, Argonne National Laboratory (1962).
4. Becker, M., *The Principles and Applications of Variational Methods*, M.I.T. Press, Cambridge, Mass. (1964), Chap. 3.
5. Becker, M., *ibid.*, Chap. 4.,
6. Calame, G. P., and F. D. Federighi, "Variational Approach to Thermal Group Calculations," *Naval Reactors Physics Handbook*, I, USAEC, Washington (1964), p. 478.
7. Calame, G. P., and F. D. Federighi, "A Variational Procedure for Determining Spatially Dependent Thermal Spectra," *Nucl. Sci. Eng.*, **10**, 190 (1961).
8. Clark, M., and K. F. Hansen, *Numerical Methods of Reactor Analysis*, Academic Press, New York (1964), p. 204.
9. Clark, M., and K. F. Hansen, *ibid.*, pp. 87–145.
10. Clark, M., and K. F. Hansen, *ibid.*, p. 175.
11. Clark, M., and K. F. Hansen, *ibid.*, p. 200.
12. Davey, W. G., "An Analysis of 23 ZPR-III Fast Reactor Critical Experiments," *Nucl. Sci. Eng.*, **19**, 259 (1964).
13. Davey, W. G., "*k* Calculations for 22 ZPR-III Assemblies Using ANL Cross Section Set 635," ANL-6570, Argonne National Laboratory (1962).
14. Davison, B., *Neutron Transport Theory*, Oxford at the Clarendon Press, London (1958), p. 21.
15. Davison, B., *ibid.*, p. 97.
16. Dirac, P. A. M., *The Principles of Quantum Mechanics*, 4th ed., Oxford at the Clarendon Press, London (1958).

17. Driscoll, M. J., "Characterization of Neutron Spectra in Fast Reactors," Sc.D. Thesis, Department of Nuclear Engineering, M.I.T. (1965).
18. Driscoll, M. J., and I. Kaplan, "A Simple Shape Function for Correlating Fast Reactor Spectra," *Trans. Am. Nuc. Soc.*, **9**, 137 (1966).
19. Dougherty, D. E., and C. N. Shen, "The Space-Time Neutron Kinetic Equations Obtained by the Semi-direct Variational Method," *Nucl. Sci. Eng.*, **13**, 141 (1962).
20. Engle, L. B., *et al.*, "Reactivity Contributions of Various Materials in Topsy, GODIVA, and JEZEBEL," *Nucl. Sci. Eng.*, **8**, 543 (1960).
21. Friedman, B., *Principles and Techniques of Applied Mathematics*, John Wiley & Sons, Inc., New York (1956).
22. Gasildo, J. M., *et al.*, "Critical Studies of a Dilute Fast Reactor Core," ANL-6338, Argonne National Laboratory (1961).
23. Glasstone, S., and M. C. Edlund, *Nuclear Reactor Theory*, D. Van Nostrand Co., Inc., New York (1952), pp. 172–189.
24. Goldstein, H., "Statistical Model Theory of Neutron Reactions and Scattering," *Fast Neutron Physics*, II, Interscience Publishers, New York (1963), p. 1525.
25. Guernsey, J. B., and D. A. Lind, "Excitation Functions for Neutron Inelastic Scattering," *Fast Neutron Physics*, II, Interscience Publishers, New York (1963), p. 1483.
26. Hansen, G. E., "Properties of Elementary Fast Neutron Critical Assemblies," *2nd UN Conf. PUAE*, **12**, United Nations, New York (1958), p. 84.
27. Hansen, G. E., "Status of Computational and Experimental Correlations for Los Alamos Fast Neutron Critical Assemblies," *Physics of Fast and Intermediate Reactors*, I, IAEA, Vienna (1962), p. 445.
28. Hansen, G. E., and W. H. Roach, "Six and Sixteen Group Cross Sections for Fast and Intermediate Assemblies," LAMS-2543, Los Alamos Scientific Laboratory (1961).
29. Huber, R. J., *et al.*, "Critical Studies of a Dilute Carbide Fast Reactor Core," ANL-6401, Argonne National Laboratory (1961).
30. Hurwitz, H., "Physical Interpretation of the Adjoint Flux: Iterated Fission Probability," *Naval Reactors Physics Handbook*, I, USAEC, Washington (1964), p. 864.
31. "Industry Assesses Next Stage in Power Reactor Development," *Nucleonics*, **23**, No. 1, 21 (1965).
32. Jolley, L. B. W., *Summation of Series*, Dover Publications, Inc., New York (1961), pp. 2, 8.
33. Kantorovich, L. V., and V. I. Krylov, *Approximate Methods in Higher Analysis*, Interscience Publishers, New York (1958), pp. 240–357.
34. Kaplan, S., "Synthesis Methods in Reactor Analysis," *Advances in Nuclear Science and Technology*, Vol. III, Academic Press, Inc., New York (1965).
35. Kaplan, S., "Variational Synthesis," *Naval Reactor Physics Handbook*, I, USAEC, Washington (1964), p. 710.
36. Kaplan, S., O. J. Marlowe, and J. Bewick, "Application of Synthesis Techniques to Problems Involving Time Dependence," *Nucl. Sci. Eng.*, **18**, 163 (1964).
37. Lewins, J., *Importance: The Adjoint Function*, Pergamon Press Ltd., London (1965).

38. Long, J. K., *et al.*, "Fast Neutron Power Reactor Studies with ZPR-III," *2nd UN Conf. PUAE*, **12**, United Nations, New York (1958), p. 119.
39. Meghreblian, R. V., and D. K. Holmes, *Reactor Analysis*, McGraw-Hill Book Company, New York (1960), p. 456.
40. Meneghetti, D., and J. R. White, "Recent Calculations for ZPR-III Fast Assemblies," ANL-7010, Argonne National Laboratory (1965), p. 111.
41. Messiah, A., *Quantum Mechanics*, Vol. I, Chap. 7, John Wiley & Sons, Inc., New York (1961).
42. Mikhlin, S. G., *Variational Methods in Mathematical Physics*, Macmillan, New York (1964), p. 491.
43. Mikhlin, S. G., *ibid.*, p. 74.
44. Mikhlin, S. G., *ibid.*, p. 218.
45. Mikhlin, S. G., *Integral Equations*, 2nd ed., Pergamon Press, Inc., New York (1964), p. 7.
46. Mikhlin, S. G., *ibid.*, p. 19.
47. Morse, P. M., and H. Feshbach, *Methods of Theoretical Physics*, I, McGraw-Hill Book Company, New York (1953), p. 870.
48. Morse, P. M., and H. Feshbach, *ibid.*, Vol. II, p. 1108.
49. Morse, P. M., and H. Feshbach, *ibid*, Vol. I, pp. 870–878.
50. Murley, T. E., "A Calculation Model for Fast Reactors," Sc.D. Thesis, Department of Nuclear Engineering, M.I.T. (1965).
51. Okrent, D., R. Avery, and H. Hummel, "A Survey of the Theoretical and Experimental Aspects of Fast Reactor Physics," *1st UN Conf. PUAE*, **5**, United Nations, New York (1956), p. 347.
52. Paik, N. C., "Application of the Spatial Expansion Method to the Solution of the Boltzmann Equation in Three Dimensions," Ph.D. Thesis, Department of Nuclear Engineering, M.I.T. (1965).
53. Pendleberry, E. D., and L. H. Underhill, "Validity of Transport Approximation," *Physics of Fast and Intermediate Reactors*, II, IAEA, Vienna (1962), p. 73.
54. Pomraning, G. C., "A Numerical Study of the Method of Weighted Residuals," *Nucl. Sci Eng.*, **24**, 291 (1966).
55. Pomraning, G. C., "An Approach to Space-Energy Problems," *Nukleonik*, **7**, 192 (1965).
56. Pomraning, G. C., "Approximate Methods of Solution of the Monoenergetic Boltzmann Equation," Ph.D. Thesis, Department of Nuclear Engineering, M.I.T. (1962).
57. Rusch, G. K., *et al.*, "Investigations of a 600-L Uranium Carbide Core," ANL-7010, Argonne National Laboratory (1965), p. 91.
58. Selengut, D. S., "Variational Analysis of a Multidimensional System," HW-59126, Hanford Laboratories (1959), p. 89.
59. Soodak, H., "Theory of Neutron Transport," *Reactor Handbook*, Vol. III-A, Interscience Publishers, New York (1962), pp. 108–110.
60. Soodak, H., *ibid.*, p. 124.
61. Stewart, L., "Leakage Neutron Spectrum from a Bare Pu-239 Assembly," *Nucl. Sci. Eng.*, **8**, 595 (1960).
62. Szwarcbaum, G., M. Sieger, and S. Yiftah, "Inelastic Scattering of High Energy Neutrons in Fast Reactors," *3rd UN Conf. PUAE*, **6**, United Nations, New York (1965), p. 189.

63. Terrell, J., "Fission Neutron Spectra and Nuclear Temperatures," *Phys. Rev.*, **113**, 527 (1959).
64. Varshni, Y. P., "Comparison of Three Equations for the Spacings of Nuclear Levels," *Nuovo Cimento*, **22**, 145 (1961).
65. Wachspress, E. L., and M. Becker, "Variational Multichannel Synthesis with Discontinuous Trial Functions," KAPL-3095, Knolls Atomic Power Laboratory (1965).
66. Weinberg, A. M., and E. P. Wigner, *The Physical Theory of Neutron Chain Reactors*, Univ. of Chicago Press, Chicago (1958), p. 223.
67. Zelazny, R., and A. Kuszell, "Multigroup Neutron Transport Theory," *Physics of Fast and Intermediate Reactors*, II, IAEA, Vienna (1962), p. 55.
68. Yiftah, S., D. Okrent, and P. A. Moldauer, *Fast Reactor Cross Sections*, Pergamon Press, Inc., New York (1960).

# Index